BIORENEWABLE RESOURCES

RESOURCES

Engineering New Products from Agriculture

BIORENEWABLE RESOURCES

Engineering New Products from Agriculture

ROBERT C. BROWN

Iowa State Press
A Blackwell Publishing Company

Robert C. Brown received his MS and PhD degrees in mechanical engineering from Michigan State University, East Lansing. He holds the Bergles Professorship in Thermal Sciences at Iowa State University where he is Professor of mechanical engineering and chemical engineering and the Director of the Center for Sustainable Environmental Technologies at Iowa State University, Ames. He is a Fellow of the American Society of Mechanical Engineers.

© 2003 Iowa State Press
A Blackwell Publishing Company
All rights reserved

Iowa State Press
2121 State Avenue, Ames, Iowa 50014

Orders: 1-800-862-6657
Office: 1-515-292-0140
Fax: 1-515-292-3348
Web site: www.iowastatepress.com

♾ Printed on acid-free paper in the United States of America

First edition, 2003

Library of Congress Cataloging-in-Publication Data

Brown, Robert C.
  Biorenewable resources: engineering new products from agriculture
Robert C. Brown—1st ed.
    p. cm.
Includes bibliographical references and index.
  ISBN 0-8138-2263-7 (alk. paper)
  1. Biomass energy. I. Title.
  TP339.B76 2003
662'.88—dc21

2002154437

The last digit is the print number: 9 8 7 6 5 4 3 2 1

Contents

Preface

This book was written as a text for upper level undergraduate students and first year graduate students in science and engineering who are seeking a broad perspective of the emerging field of biorenewable resources. By its very nature, the conversion of biorenewable resources, also known as biomass, into biobased products and bioenergy requires a system perspective. Traditional academic disciplines have been organized to provide students in-depth training and intellectual focus in a single field such as agriculture, chemistry, engineering, environmental science, or economics. Students usually gain broader perspectives after graduation, upon employment in established industries where commercially successful enterprises are based on systems that encompass many disciplines. However, there is growing recognition that new employees in any industry would be more immediately productive if they started with a broader view of the world. This is particularly true for work in biorenewable resources, an industry that is not widely deployed or mature.

The ten chapters of this book do not assume any previous training in biorenewable resources, although most students should have undergraduate training in science or engineering. Chapter 1 is an introduction to the field of biorenewable resources, which includes a brief history of the use of biorenewable resources and a description of the motivations for advancing the biobased products industry. Chapter 2 provides fundamental concepts for understanding the processing of biorenewable resources: engineering thermodynamics, organic chemistry, and plant chemistry. This chapter is aimed at students who may have deficiencies in these concepts or who desire a review of the topics. Chapter 3 is a description of biorenewable resources. This chapter includes sections that define the resource base, describes properties that are important to the handling and processing of biorenewable resources, provides information on yields of various kinds of biomass, and assesses the availability of different kinds of biorenewable resources, including waste biomass and dedicated energy crops. Chapter 4 is an introduction to production of biorenewable resources. Separate sections are devoted to herbaceous crops and woody crops. This chapter also includes descriptions of storage systems and the prospects for using transgenic crops in production of biorenewable resources. Chapter 5 is an introduction to the wide array of products that are currently produced or anticipated from biorenewable resources. Major topics in this chapter include bioenergy, transportation fuels, chemicals, and natural fibers.

The next three chapters are devoted to the processes by which biorenewable resources are transformed into biobased products and bioenergy. Chapter 6 details technologies for converting the chemical energy of plant materials into thermal energy and gaseous fuels suitable for process heat and stationary power. Chapter 7 describes technologies for processing biorenewable resources into chemicals with applications ranging from transportation fuels to the manufacture of consumer products. Chapter 8 explains how natural fibers can be separated from biorenewable resources for use in the manufacture of paper and building materials.

The final two chapters deal with environmental and economic issues. Chapter 9 describes the environmental impact of producing and processing biorenewable resources and using the resulting products. This chapter also describes environmental concerns associated with the use of transgenic crops as biorenewable resources. Chapter 10 introduces students to the economics of biorenewable resources. The chapter includes separate discussions on estimating the costs of producing crops and manufacturing biobased products. The chapter concludes with specific cost estimates for various biobased products.

I would like to acknowledge several people who influenced the course of this book long before it was conceived. Ed Woolsey hustled me into the biomass business over a decade ago when he convinced me to evaluate an old "Buck Rogers" gasifier. A few years later, Joel Snow expanded my parochial view of biomass to encompass both energy and products when, as director of the Institute for Physical Research and Technology, he asked me to develop a research program in biorenewable resources for Iowa State University. Floyd Barwig and Norm Olson at the Iowa Energy Center provided generous resources to both study and write about the subject during the last several years. I am grateful to Judi Brown at Iowa State Press for suggesting the book. The resourcefulness and efficiency of Diane Love and Tonia McCarley at the Center for Sustainable Environmental Technologies made it possible for me to devote time to writing this book in the midst of my other pressing responsibilities.

I am indebted to several colleagues for reading chapters and making suggestions, with special thanks to Forest Brown, Janet Cushman, Maohong Fan, Robert McCarley, Graeme Quick, Brent Shanks, and Jerod Smeenk. Several graduate students were kind enough to read the manuscript cover to cover: Nathan Brown, Keith Cummer, Daren Daugaard, Nathan Emsick, and Dave Falkowski. Of course, errors and omissions are solely my responsibility.

I am especially grateful to my family, not only for their patient tolerance when I cloistered myself evenings and weekends to write, but for making the book a family affair: Carolyn prepared the line drawings; Tristan was my able research assistant; Trevor managed my computer problems; and Daniel provided baseball games on Sunday afternoons, which were welcome breaks from writing.

BIORENEWABLE RESOURCES

RESOURCES

Engineering New Products from Agriculture

Introduction

1.1 Transitions

Humankind has gone outside the biotic environment for the majority of its material needs only recently. Plant-based resources were the predominant source of energy, organic chemicals, and fibers in the West as recently as 150 years ago, and they continue to play important roles in many developing countries. The transition to non-biological sources of energy and materials was relatively swift and recent in the history of the world. Many assumed that the transition was irreversible. Indeed, in the 1960s it appeared fossil fuels were only a short bridge to the nuclear age: fission reactors would supply energy for power and transportation fuels in the waning decades of the twentieth century, while fusion reactors would ultimately provide limitless supplies of energy. In this scenario, petroleum and natural gas would continue as the source of building blocks for organic chemical synthesis.

Reality has proved more complicated, uncertain, and unsettling. In the United States, nuclear fission has not lived up to its promise and has become a political pariah in the face of reactor safety concerns and unresolved questions about radioactive waste disposal. Nuclear fusion has a lively history of tokamaks, inertial confinement, and cold fusion, but no breakeven in energy output. Petroleum continues to supply most of the world demand for transportation fuels and commodity chemicals, but the remaining reserves are increasingly concentrated in the hands of a few capricious nations. Coal is plentiful but introduces tremendous burdens on the environment ranging from acid rain to mercury poisoning. The hypothesized global impact of greenhouse warming due to carbon dioxide emissions raises questions about the future use of all fossil fuels. In this political and social climate, there are calls for development of reliable and long-term resources that have fewer environmental impacts than fossil resources. This book explores a return to the biotic environment for energy, chemicals, fuels, and fibers.

1.2 Definitions

Biorenewable resources, also known as biomass, are organic materials of recent biological origin. Biomass may be grown as crops, but the vast majority of the world's biorenewable resources are forests, prairies, marshes, and fisheries. The energy to build the chemical bonds of these organic materials comes from sunlight. Solar energy collected by green plants is converted into energetic chemical bonds to produce proteins, oils, and carbohydrates. This stored chemical energy is raw material that can be used as a resource for heat and stationary power, transportation fuels, commodity chemicals, and fibers. In contrast, coal, oil, and natural gas are thought to be derived from plants and animals buried and transformed into fossil fuels eons ago. In this view, the only difference between biomass and fossil fuels is the degree that nature processes biogenic materials (that is, those derived from living organisms).[1]

   Biorenewable resources are by definition sustainable natural resources. Sustainable implies that the resource renews itself at such a rate that it will be available for use by future generations. Thus, a stand of grass cut every year for a hay crop represents a biorenewable resource while a field cleared in a tropical rainforest for "slash and burn" agriculture is not. Biorenewable resources can be converted into either bioenergy or biobased products.

   Bioenergy, also known as biomass energy, is the conversion of the chemical energy of a biorenewable resource into heat and stationary power. The process might be as simple as burning logs in a fireplace to warm a room or as complex as gasifying plant fibers to a mixture of hydrogen and carbon monoxide, removing contaminants, and electrochemically converting the gaseous fuel to electricity within a high-temperature fuel cell.

   Biobased products include transportation fuels, chemicals, and natural fibers derived from biorenewable resources. *Transportation fuels* are generally liquid fuels, such as ethanol or biodiesel, but compressed hydrogen and methane have also been proposed and evaluated for use in vehicular propulsion. These fuels can be considered a form of bioenergy because their ultimate purpose is to produce mechanical power. However, since the markets for electric power and transportation are distinct, these two energy products are best segregated into the categories of bioenergy and transportation fuels, respectively. *Chemicals* may include pharmaceuticals, nutraceuticals, and other fine chemicals, but the emphasis here is on commodity chemicals, which produce high demands for biorenewable resources. An example is polylactic acid, which is derived from the fermentation of sugars hydrolyzed from cornstarch and can be converted into biobased polymers used in a variety of consumer products such as carpets. *Natural fibers* are bundles of long, thin plant cells with durable walls of lignocellulose. (Animal fibers are also natural fibers but are not treated in this book.) We will distinguish between natural fibers, which occur naturally in plant tissues, and synthetic fibers manufactured from polymers, regardless of whether the source of the polymers is petrochemicals or biobased chemicals.

1.3 Brief History of Biorenewable Resource Utilization

Many of the advances in early human society were based on exploitation of biorenewable resources. The first campfire, dating to as early as 500,000 B.P. (before present), was most surely kindled from wood rather than coal. Evidence that vegetable oils and animal fats were sources of illumination by 40,000 years B.P. is found at Upper Paleolithic cave sites in Europe. Draft animals represented mankind's first use of prime movers other than their own muscle power. Grasses and cultivated grains, fed to draft animals, provided power and transportation needs in early societies.

Except for stone tools, virtually all possessions until the advent of the Bronze Age were biobased products. Wood was a versatile composite material for construction of hunting and farming implements. Cotton fibers were spun and twined in Peru as long ago as 12,000 B.P. Fermentation of sugars to ethanol was mastered as long ago as 6,000 B.P., but the product remained too precious for any but convivial purposes for thousands of years afterward.

Wood was the primary fuel for most pre-twentieth century societies, although there are traditions of grass-fueled steel making in Africa and using "lightly processed" grass in the form of dried buffalo dung for cooking fires in the American West. The preeminence of bioenergy began to wane when the great forests of the Northern Hemisphere were depleted by the voracious fuel demand in the manufacture of copper, iron, and glass, and the powering of steam engines. By the mid-eighteenth century, coal had supplanted wood as the primary energy source for European and North American countries.

Biorenewable resources continued to be important as sources of chemicals and materials for another seventy-five years, aided by advances in industrial chemistry. Gum turpentine, rosins, and pitches, collectively known as naval stores, were extracted from coniferous trees for maintaining wooden ships and ropes. Natural latex from the hevea rubber tree was vulcanized to an elastic, waterproof material used in numerous consumer products. The "destructive distillation" of wood yielded methanol (wood alcohol) and other industrially important compounds. Wood pulping was developed to separate cellulose fibers used in paper and cardboard products. From cellulose came the first semi-synthetic plastic, celluloid. Advances in the brewery industry eventually led to commercial fermentation of a variety of organic alcohols and acids with applications far beyond the beverage industry.

With the exceptions of lumber production for building materials, fiber production (wood pulping and cloth manufacture), and ethanol fermentation, the manufacture of biobased products rapidly declined in the twentieth century. This decline was not the result of resource scarcity, as was the case for bioenergy. Instead, rapid advances in the chemistry of coal-derived compounds during the late nineteenth century, followed by the development of the petrochemical industry in the early twentieth century, provided less expensive and more easily manipulated feedstock for the production of chemicals and materials.

1.4 Motivation for a Return to Bioenergy and Biobased Products

Economics currently do not favor bioenergy and biobased products. However, other factors are coming into play that increase the attractiveness of biorenewable resource utilization. These factors include desires to improve *environmental quality*, concerns that *national security* is compromised by over reliance on foreign sources of fossil fuels, the presence of *excess agricultural production* capacity in many developed nations, and the importance of *rural development* in improving economies of many agricultural regions.

Environmental Quality

The production and utilization of fossil fuels introduce several environmental burdens of increasing concern. Production includes mining for coal and drilling for petroleum and natural gas. Most of the impacts are local in nature. Mining leaves behind spoil piles and acid drainage. Drilling can result in oil spills or sites contaminated by drilling mud or brackish water.

Utilization of fossil fuels can yield a plethora of *local, regional,* and *global impacts*. *Local impacts* can arise from solid waste disposal sites for ash from coal combustion, or coke or sulfur from petroleum refineries. Combustion of coal or petroleum-based motor fuel also produces carbon monoxide, fine particulate matter, and smog as local pollution problems. *Regional impacts* of fossil energy utilization are mostly the result of acid rain, which is generated from sulfur dioxide and nitrogen oxides released during combustion. Acid rain can affect environments half a continent away from the point of pollutant emission. *Global impacts* are of two kinds: ozone depletion in the stratosphere and global climate change. Ozone depletion is usually associated with the release of chlorofluorocarbons from refrigeration equipment and aerosol cans. Nitrogen oxides have also been implicated in upper atmosphere reactions that destroy ozone molecules. However, the greatest concern with regard to global impact is the possible role of carbon dioxide released during combustion in contributing to the greenhouse effect in the atmosphere. Although the magnitude of this problem is not known at present, there have been calls to greatly reduce the net rate of emission of carbon dioxide from the use of fossil fuels.

Bioenergy and biobased products are not a panacea for these problems. However, the environmental burden from the use of biorenewable resources is generally much less than from the use of fossil resources. An exact accounting of the benefits of using biorenewable resources is a difficult and sometimes politically charged process. For example, some argue that the benefit of using a 10% blend of ethanol in gasoline to reduce carbon monoxide emissions from spark-ignition engines is outweighed by the increased release of hydrocarbons (a factor in smog formation) into the atmosphere due to the greater volatility of this fuel blend. These issues are addressed in detail in Chapter 9.

National Security

In 1974, a severe economic crisis developed in many parts of the world as a result of disruptions in the distribution of petroleum to markets. The so-called Energy Crisis was commonly ascribed to dwindling reserves of petroleum resources, and many experts predicted that looming energy scarcity would drive petroleum prices from a few dollars to over $100 per barrel, plunging the world into an economic depression that would be difficult to reverse. These concerns provided the impetus for a short-lived effort in the United States to commercially implement alternative energy sources, including solar, wind, biomass, and coal. However, by 1980 petroleum prices had considerably moderated and it was understood that an actual shortage of energy had never existed. Instead, a decision by Arab nations to boycott the sale of petroleum to the United States was responsible for temporary escalations in the world price of oil. The boycott was an effort to influence U.S. foreign policy in the Middle East, a policy that favored the state of Israel in its conflicts with surrounding Arab states. Although disruptive in the short-term, ultimately the boycott and various production quotas were lifted and the world price for petroleum dropped. This threat to national security failed because the United States and other nations responded by reducing their dependence on foreign sources of oil through energy conservation and energy efficiency improvements, and by switching to domestic energy sources, mostly coal, natural gas, and petroleum. As demand for Middle Eastern petroleum dropped these countries saw the major source of their revenues evaporating; economic survival forced them to suspend their efforts to use "the oil weapon," as it was called.

The lesson learned from the "Energy Crisis" was that effective national security incorporates an energy policy that reduces heavy reliance on foreign cartels for energy resources. However, there is some evidence that this lesson is not being heeded in the United States, where dependence on petroleum imports now exceeds that of 1974. As a percentage of petroleum demand, petroleum imported into the United States exceeded 50% in 2001 and is expected to grow to 64% by 2020. With two-thirds of world petroleum reserves located in the Middle East (including the Caspian basin), increased dependence on imported sources is inevitable unless domestic energy sources such as biorenewable resources are developed.

Excess Agricultural Production

A frequently expressed concern about shifting agriculture toward production of biobased products is the potential impact on food production. Securing a safe and inexpensive food supply is the keystone of agricultural policy in the United States. Many people oppose the use of agricultural lands for the production of bioenergy and biobased products on the grounds that, at best, domestic food prices will rise and, at worse, starvation will increase in developing countries that are dependent on agricultural imports.

In fact, agricultural production in excess of domestic use and export demands exists in the United States as well as in a growing number of other countries. The United States in 1990, with a population of 250 million people, had 12% less land in agricultural production than it did in 1929, when the population was only 120 million. The reason for this decline in agricultural lands is primarily due to increasing crop productivity. For example, U.S. corn yields between 1929 and 1990 increased from 22 to 30 bushels per acre to 101–139 bushels per acre. These improvements are the result of advances in plant genetics, fertilizers, pesticides, and production practices. In an effort to keep production in balance with demand, the U.S. government encouraged development of export markets for agricultural products in the last half of the twentieth century. However, this strategy did not adequately anticipate the developing world's ability to feed itself. Even China, often viewed as an unlikely candidate for self-sufficiency because of its burgeoning population of over 1 billion people, has in recent years become a net exporter of agricultural products.

Recognizing that overproduction threatens the stability of agriculture, the U.S. government instituted a program called the Conservation Reserve Program to deliberately remove marginal lands from production. Producers are allowed to grow grasses or trees on enrolled acreage but cannot harvest this biomass. Almost 34 million acres were enrolled in this program in 2001 compared to about 900 million acres left in production. As described in Chapter 3, millions of additional acres are currently devoted to low productivity uses such as pasturage or woodlots that could provide additional land for production of feedstocks for biobased products. Thus, devoting some land to biorenewable resources poses no immediate threat to food prices in the United States nor to feeding the developing countries of the world.

Rural Development

The impact of modern agriculture has not been completely positive. Increases in labor productivity in agriculture have reduced labor costs but have also contributed to the depopulation of rural communities. Improvements in transportation have expanded opportunities for trade but in turn have put U.S. farmers in direct competition with producers in developing nations, where land values are a fraction of what they are in the U.S. The highly integrated agricultural processing industry is successful in capturing value from agriculture, but little of that value is shared by the producers: while return on investment in the food processing industry is typically 15%, production agriculture rarely yields returns greater than 1–3%.

As a result, agriculture in developed nations is increasingly dependent on government subsidies to be viable. In the United States government payments represent in excess of 50% of gross income for typical producers. Farm subsidies in 1997 reached $23 billion. Despite this assistance producers are increasingly turning to off-farm jobs for supplemental income to support their families.

Both producers and rural communities are looking for new opportunities to boost income and economic development. Development of crops for new markets, especially crops that are processed locally into value-added products, would provide significant opportunities for rural development.

1.5 Challenges in Using Biorenewable Resources

Biorenewable resources have a number of disadvantages compared to the fossil resources with which they compete, including the fact that most biorenewable resources are solid materials of low bulk density, high moisture content, low heating value, and high oxygen content.

The solid nature of biomass is both a blessing and a curse. Solids are easily collected by hand and can be stored in bins. These advantages were especially important to early human societies, which did not have technology to handle large quantities of liquids and gases. However, solids are notoriously difficulty to handle and process in the automated industries of the modern world.

Gases and liquids can be moved hundreds and even thousands of miles through pipelines and stored in tanks with a minimum of human intervention. These fluids do not clog pumps and pipes unless they are carrying solids along with them. The flow of gases and liquids is easily metered with relatively simple instruments, and their flow rates regulated by the turn of a valve. Gases and liquids can be rapidly and easily dispersed or mixed, which allows them to be readily processed into heat, power, fuels, and chemicals.

Monolithic solids, of course, do not have the property of flow that expedites the handling of gases and solids. However, if solids are broken up into small particles and aerated, they acquire flow properties resembling those of liquids, a fact that has been widely exploited in the twentieth century. Grain is sucked from the holds of ships by giant vacuum cleaner-like devices rather than being carried up in 50-kg sacks on the backs of stevedores. Coal is blown as a fine powder into giant steam boilers rather than being shoveled onto grates by firemen. Even the backbreaking work of shoveling snow from driveways has been eliminated by snow blowers that chop up the snow and convey it pneumatically.

Nevertheless, the handling of solids remains a problem. The process of converting solids into granular or powdered materials is an energy intensive process: as little as 10% of the energy consumed by crushing, grinding, or cutting machines actually goes into dividing the materials, while the rest is dissipated as thermal energy. Particulate materials do not flow as smoothly or predictably as gases and liquids. Particles as coarse as woodchips or as fine as flour can easily clog hoppers and transport lines. Solids-metering and control devices are neither as reliable, nor as available, as devices for gas and liquid flows. Finely divided solids can present special erosion problems and explosion hazards. Solids-handling systems are uniformly acknowledged as high maintenance items in industrial processes.

Typically, the density of biomass is so low that the volume rather than the weight of biomass that can be transported will limit the capacity of the transportation systems. Accordingly, the number of trucks or railcars required to supply a conversion facility will increase as the volumetric density of the fuel decreases. For example, a conventional steam power plant with a relatively modest electric power output of 50 MW would require up to 75 tractor-trailer loads of biomass per day to stoke the boilers, compared to as few as 28 weight-limited loads of coal. Similar arguments apply to the size of the boiler. A boiler designed to burn biomass must be much larger than a coal-fired boiler of comparable thermal output because of the lower energy density of biomass compared to coal.

The moisture content of green biomass also detracts from its performance as fuel. Freshly harvested biomass can have moisture content of 50% or more. This additional weight adds needlessly to the cost of transporting the fuel to a conversion facility. In some conversion processes, such as anaerobic digestion, high moisture content may be beneficial to the conversion process. In other cases, such as direct combustion, high moisture exacts a high penalty on the conversion process. Field drying is feasible for many biomass crops, but may only reduce moisture content to 20–25%. For some processes additional drying at the conversion plant is required. Freshly mined coal also contains moisture: in some western coals it can be as high as 30%. However, moisture is generally much less of a problem in coal.

Biomass must compete with a variety of fossil resources, including petroleum, natural gas, and coal. Since they are both solid fuels, substituting biomass for coal might seem a more competitive entry into energy markets dominated by fossil fuels. However, on a purely thermodynamic basis, biomass is generally inferior to coal. Coals have heating values typically in the range of 23–28 MJ/kg. On a mass basis, the heating values of biomass are 16–20 MJ/kg, which are 20–30% lower than coal. Exacerbating this situation are the significantly lower densities of biomass compared to coal: mined coal has a bulk density of around 880 kg/m^3 compared to 545 kg/m^3 for hybrid poplar logs and 230 kg/m^3 for baled switchgrass. On a volumetric basis, the heating value of biomass is only 20–50% that of coal. Volumetric heating value is an important consideration in both the transport of biomass to a conversion facility and in the conversion process itself.

Most biorenewable resources contain a significant portion of oxygen—up to 45%-wt for carbohydrates. In contrast, fossil resources contain substantially less oxygen—as high as 25%-wt in lignite coal and virtually absent in natural gas and petroleum. Although "oxygenated" fuels are touted for their environmental performance as motor fuels, in general, chemically bonded oxygen is responsible for the lower heating values of biobased fuels as well as many of the difficulties of substituting biobased chemicals for petroleum-based chemicals. In principle, oxygenated compounds derived from biorenewable resources can be converted to hydrocarbons, but the process is relatively expensive and cannot currently compete economically with petroleum-derived hydrocarbons.

1.6 Foundations for a Biorenewable Resources Economy

An economy built upon biorenewable resources must be able to supply transportation fuels, commodity chemicals, natural fibers for fabrics and papermaking, and energy for process heat and electric power generation. There is precedence for producing all of these products from biorenewable resources. However, many of the conversion technologies currently do not yield products that are cost-competitive with the fossil-based products that dominate today's markets. This situation is likely to change as technologies improve for producing biobased products, and environmental and political factors make green products from indigenous resources more attractive. The remaining chapters of this book explore various opportunities for making biobased products.

Note

1. For a fascinating presentation of an abiogenic hypothesis on the origin of fossil fuels, see Thomas Gold's "The Deep Hot Biosphere."

Further Reading

Eaglesham, A., W. F. Brown and R. W. F. Hardy, eds. 2000. *The Biobased Economy of the Twenty-First Century: Agriculture Expanding into Health, Energy, Chemicals, and Materials.* NABC Report 12. New York: National Agricultural Biotechnology Council.

National Research Council. 2000. *Biobased Industrial Products: Priorities for Research and Commercialization.* Washington, D.C.: National Academy Press.

Smil, V. 1994. *Energy in World History.* Boulder: Westview Press.

Problems

1.1 Define biorenewable resources. List five examples of different kinds of biorenewable resources.

1.2 What was the major factor in the rapid decline in the use of biobased products in the twentieth century in favor of petroleum-based products?

1.3 List the four motivations for moving from an economy based on fossil resources to one based on biorenewable resources.

1.4 One argument against the use of biorenewable resources for production of biobased products and bioenergy is concern that food supplies would be adversely affected. What factors mitigate this concern?

1.5 What are some of the difficulties in working with solid biomass compared to liquid and gaseous hydrocarbons?

Fundamental Concepts in Understanding Bioenergy and Biobased Products

2.1 Introduction

The foundations of bioenergy and biobased products are engineering thermodynamics, organic chemistry, and plant chemistry. This chapter is designed to introduce or reacquaint readers, as appropriate, to fundamental concepts in these three disciplines. The treatments do not pretend to be exhaustive; readers requiring additional background are directed to the list of reference materials at the end of this chapter.

2.2 Engineering Thermodynamics

Mass and energy balances on process streams are critical to understanding the efficiency and economics of bioenergy. Accounting for these balances is more complicated than for energy conversion processes that do not include chemical reaction because chemical constituents change and energy is released from the rearrangement of chemical bonds.

2.2.1 Mass Balances

In the absence of chemical reaction, the change in mass of a particular constituent within a control volume is equal to the difference in net mass flow of the constituent entering and exiting the control volume. Figure 2.1 illustrates mass balance for a system consisting of five inlets and five exits. In general, the mass balance for a given chemical constituent can be written in the form:

$$\frac{dm_{CV}}{dt} = \sum_i \dot{m}_i - \sum_e \dot{m}_e \qquad (2.1)$$

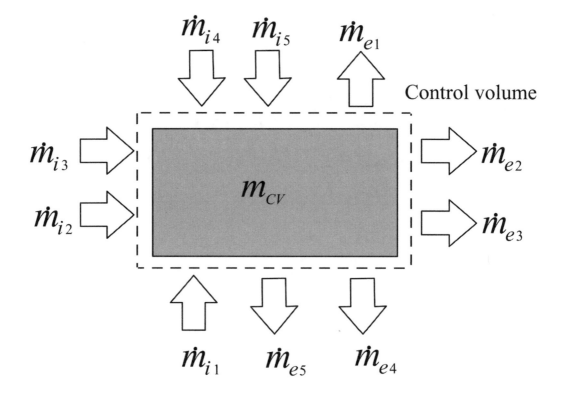

FIG. 2.1. Mass balance on steady-flow control volume with five inlets and five exits.

where m_{CV} is the amount of mass contained within the control volume, $\dot{m}_i$ and $\dot{m}_e$ are, respectively, the rates at which mass enters at i and exists at e, where we allow for the possibility of several inlets and exits. For steady flow conditions, the net quantity of mass in the control volume is unchanging with time and Eq. 2.1 can be written as:

$$\sum_i \dot{m}_i = \sum_e \dot{m}_e \qquad (2.2)$$

However, when chemical reaction occurs, chemical compounds are not conserved as they flow through the system. For example, methane (CH_4) and oxygen (O_2) entering a combustor are consumed and replaced by carbon dioxide (CO_2) and water (H_2O):

$$CH_4 + 2O_2 \rightarrow CO_2 + 2H_2O \qquad (2.3)$$

Accordingly, mass balances cannot be written for methane and oxygen using either Eq. 2.1 (unsteady flow) or Eq. 2.2 (steady flow). Although chemical compounds are not conserved, the chemical elements making up these compounds are conserved; thus, elemental mass balances can be written.

In the case of the reaction of CH_4 with O_2, mass balances can be written for the chemical elements carbon (C), hydrogen (H), and oxygen (O). However, because chemical compounds react in distinct molar proportions, it is usually more convenient to write *molar* balances on the elements. Recall that a mole of any substance is the amount of mass of that substance that contains as many individual entities (whether atoms, molecules, or other particles) as there are atoms in 12 mass units of carbon-12. For engineering systems it is usually more convenient to work with kilograms as the unit of mass; thus, for this measure kilomole (kmol) will be employed instead of the gram-mole (gmol) that often appears in chemistry books. The number of kilomoles of a substance, n, is related to the number of kilograms of a substance, m, by its molecular weight, M:

$$n = \frac{m}{M} \tag{2.4}$$

On a molar basis, it is straightforward to account for mass changes that occur during chemical reactions: an overall chemical reaction is written that is supported by molar balances on the elements appearing in the reactant and product chemical compounds.

> **Example:** One kilogram of methane reacts with air. (a) If all of the methane is to be consumed, how many kilograms of air will be required? (b) How many kilograms of carbon dioxide, water, and nitrogen will appear in the products?
>
> One kilogram of methane, with a molecular weight of 16, is calculated to be 1/16 kmol using Eq. 2.4. Air is approximated as 79% nitrogen and 21% oxygen on a molar basis. The overall chemical reaction can be written as:
>
> $$(\tfrac{1}{16})CH_4 + a(O_2 + \tfrac{0.79}{0.21}N_2) \rightarrow xCO_2 + yH_2O + zN_2$$
>
> where a is the number of kilomoles of oxygen required to consume 1/16 kmol of CH_4 and x, y, and z are the kilomoles of CO_2, H_2O, and N_2, respectively, in the products. The unknowns in this equation can be found from molar balances on the elements C, H, O, and N:
>
> carbon: $\tfrac{1}{16} = x \, (\text{kmoles})$
>
> $\therefore m_{CO_2} = n_{CO_2} \times M_{CO_2} = \tfrac{1}{16} \times 44 = 2.75 \text{ kg}$

hydrogen: $\frac{1}{16} \times 4 = 2y$

$y = \frac{1}{8}$ (kmoles)

$\therefore m_{H_2O} = n_{H_2O} \times M_{H_2O} = \frac{1}{8} \times 18 = 2.25$ kg

oxygen: $2a = 2x + y = 2 \times \frac{1}{16} + \frac{1}{8} = \frac{1}{4}$

$a = \frac{1}{8}$ (kmoles)

$\therefore m_{O_2} = n_{O_2} \times M_{O_2} = \frac{1}{8} \times 32 = 4$ kg

nitrogen: $\frac{0.79}{0.21}a = \frac{0.79}{0.21} \times \frac{1}{8} = 0.47 = z$ (kmoles)

$\therefore m_{N_2} = n_{N_2} \times M_{N_2} = 0.47 \times 28 = 13.2$ kg

A check shows that 18.2 kg of methane and air are converted into 18.2 kg of products in the form of carbon dioxide, water, and nitrogen, as expected from mass conservation.

In some processes, like combustion and gasification, it is useful to compare the actual oxygen provided to the fuel to the amount theoretically required for complete oxidation (the stoichiometric requirement). There are several useful comparisons, described below.

The fuel-oxygen ratio, F/O, is defined as the mass of fuel per mass of oxygen consumed (a molar fuel-oxygen ratio is also sometimes defined). Another frequently used ratio is the equivalence ratio, ϕ:

$$\phi = (F/O)_{actual}/(F/O)_{stoichiometric} \tag{2.5}$$

This ratio is less than unity for fuel-lean conditions and greater than unity for fuel-rich conditions. Two other ratios, theoretical air and excess air, can be calculated on either mass or molar bases:

$$\text{Theoretical air (\%)} = (\text{actual air/stoichiometric air}) \times 100 \tag{2.6}$$

$$\text{Excess air (\%)} = \left(\frac{\text{actual air} - \text{stoichiometric air}}{\text{stoichiometric air}} \right) \times 100 \tag{2.7}$$

2.2.2 Energy Balances

In the absence of chemical reaction, the net change in stored energy within a control volume is given by the net flow of energy into the control volume in the form of heat and work as well as kinetic energy, potential energy, and enthalpy associated with mass flowing into and out of the control volume. Figure 2.2 illustrates the energy balance for a control volume with two inlets and one outlet and with work

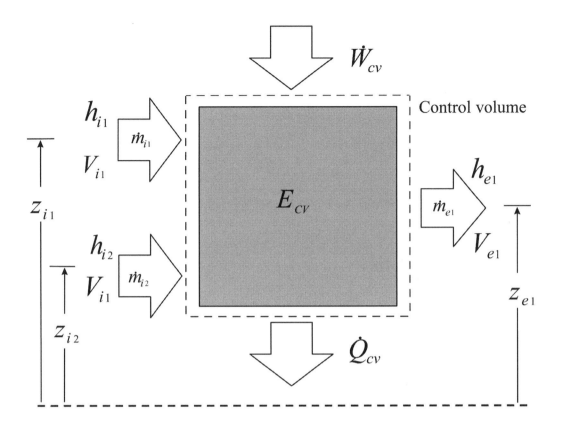

FIG. 2.2. Energy balance on steady-flow control volume with two inlets and one exit.

transferred in and heat transferred out. More generally, an open system can be described by an energy balance of the form:

$$\frac{dE_{CV}}{dt} = \dot{Q}_{CV} - \dot{W}_{CV} + \sum_i \dot{m}_i\left(h_i + \tfrac{1}{2}V_i^2 + gz_i\right) - \sum_e \dot{m}_e\left(h_e + \tfrac{1}{2}V_e^2 + gz_e\right) \qquad (2.8)$$

where E_{CV} is the stored energy in the control volume; $\dot{Q}_{CV}$ and $\dot{W}_{CV}$ are the rates at which heat and work cross the control volume boundary; h is enthalpy; V is velocity; and z is elevation with respect to an arbitrary datum for the mass flows at the inlet, i, and outlet, e. In steady flow, with a single inlet and single outlet and no velocity or elevation changes in the system, Eq. 2.8 simplifies to:

$$\dot{Q}_{CV} - \dot{W}_{CV} = \dot{m}(h_e - h_i) \qquad (2.9)$$

However, for a chemically reacting system, this formulation of an energy balance does not take into account changes in the chemical composition of the system nor the chemical energy absorbed or released during these reactions. Like mass conservation, it is more convenient to present energy conservation in a molar formulation rather than in a mass formulation when chemical reaction occurs. The intensive property enthalpy, h, with units of kJ/kg, is replaced by the intensive property molar enthalpy, $\bar{h}$, with units of kJ/kmol. Table 2.1 is an abbreviated collection of molar enthalpies of selected gases as a function of temperature. More extensive collections are available in thermodynamics textbooks (e.g., Moran and Shapiro, Wiley, 1999).

Energy conservation on a molar basis for a steady flow system consisting of one inlet for reactants r and one outlet for products p (and neglecting velocity or elevation changes) is of the form:

$$\dot{Q}_{CV} - \dot{W}_{CV} = \sum_p \dot{n}_p \bar{h}_p - \sum_r \dot{n}_r \bar{h}_r \qquad (2.10)$$

Table 2.1. Thermodynamic properties for selected gases

T (K)	Molar Enthalpies (kJ/kmol)					
	N_2	O_2	H_2O	CO	CO_2	H_2
0	0	0	0	0	0	0
298	8,669	8,682	9,904	8,669	9,364	8,468
500	14,581	14,770	16,828	14,600	17,678	14,350
1000	30,129	31,389	35,882	30,355	42,769	29,154
1500	47,073	49,292	57,999	47,517	71,078	44,738
2000	64,810	67,881	82,593	65,408	100,804	61,400
2500	82,981	87,057	108,868	83,692	131,290	78,960
3000	101,407	106,780	136,264	102,210	162,226	97,211
3250	110,690	116,827	150,272	111,534	177,822	106,545

Substance	Formula	$\bar{h}_f^o$ (kJ/kmol)
Carbon	C(s)	0
Hydrogen	H_2	0
Nitrogen	N_2	0
Oxygen	O_2	0
Carbon monoxide	CO	−110,530
Carbon dioxide	CO_2	−393,520
Water (liquid)	H_2O (l)	−285,830
Water (vapor)	H_2O (g)	−241,820
Methane	CH_4	−74,850
Methanol (liquid)	CH_3OH (l)	−238,810
Methanol (vapor)	CH_3OH (g)	−200,890
Ethanol (liquid)	C_2H_5OH (l)	−277,690
Ethanol (vapor)	C_2H_5OH (g)	−235,310

Source: M. Moran and H. Shapiro, 1999, *Fundamentals of Engineering Thermodynamics*, 4th Edition (New York: Wiley).

Note: $\bar{h}_f^o$ = enthalpy of formation at 298°K.

where $\dot{n}$ specifies the molar flow rate of a chemical constituent and the summation is over all the products p at the exit or all the reactants r at the inlet. Integrated over a finite time interval, this equation takes the form:

$$Q_{CV} - W_{CV} = \sum_p n_p \overline{h}_p - \sum_r n_r \overline{h}_r \qquad (2.11)$$

where Q_{CV} and W_{CV} are the amounts of heat and work done over a designated time interval, and n_r and n_p are the kilomoles of reactants and products, respectively, crossing the control surface in the time interval. A convenient shorthand is to designate H_p and H_r as the mixture enthalpies (kJ) of the products and reactants, respectively, and ΔH as the change in enthalpy between the products and reactants:

$$\Delta H = H_p - H_r = \sum_p n_p \overline{h}_p - \sum_r n_r \overline{h}_r \qquad (2.12)$$

An enthalpy-temperature diagram, shown in Fig. 2.3, is useful in understanding the change in enthalpy that occurs in the presence of a chemical reaction. For a given mixture of reactants, say methane and oxygen, there is a unique enthalpy-temperature relationship. Similarly, a unique enthalpy relationship exists for the

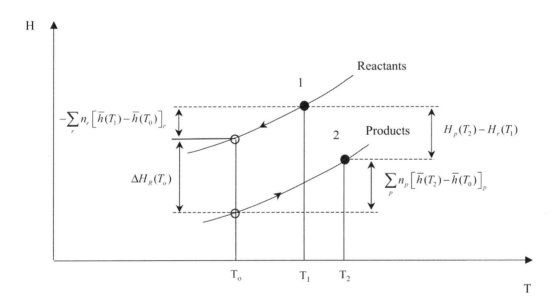

FIG. 2.3.   Relationship between mixture enthalpy and temperature for a chemically reacting system.

products of methane oxidation (carbon dioxide and water). It is also easy to under-stand the enthalpy change that occurs for a constant temperature chemical reaction: reactants at T_o are converted to products at T_o with a release or absorption of energy known as the enthalpy of reaction $\Delta H_R(T_o)$ at temperature T_o. Reactions that release energy (exothermic reactions) have negative enthalpies of reaction whereas reactions that absorb energy (endothermic reactions) have positive enthalpies of reaction. This situation becomes obvious by inspecting Eq. 2.11 and recalling the convention that Q_{cv} is negative for heat flow out of a system.

However, the typical chemical reaction is not isothermal; indeed, many combustion reactions are accompanied by temperature increases of over 1000 K. Thus, enthalpy changes must account for sensible enthalpy changes of the reactants, sensible enthalpy changes for the products, and the release or absorption of heat as a result of the chemical reaction. One way to handle this potentially complicated situation is to visualize the reaction as following the reaction pathway illustrated in Fig. 2.3: reactants initially at temperature T_1 are cooled to temperature T_o at which point the reactants undergo isothermal chemical reaction to form products that are then heated to the final temperature T_2. Thus, the enthalpy change for non-isothermal chemical reactions can be calculated from the relationship:

$$H_p(T_2) - H_r(T_1) = \sum_p n_p[\overline{h}(T_2) - \overline{h}(T_o)]_p + \Delta H_R(T_o) \qquad (2.13)$$
$$- \sum_r n_r[\overline{h}(T_1) - \overline{h}(T_o)]_r$$

where sensible enthalpies, $\overline{h}(T)$, for a variety of chemical substances are available as tabulations of thermodynamic properties of substances. Likewise, enthalpies of reaction at a specified reference temperature, T_o, have been compiled for a number of chemical reactions. In the SI system, T_o is chosen as 298 K for the purpose of tabulating data. Using Eq. 2.13, tabulations of sensible enthalpies and enthalpies of reaction can be used to calculate enthalpy changes for reactions under a wide variety of conditions.

> **Example:** One kilomole of biogas produced by anaerobic digestion of animal waste consists of 60% methane and 40% carbon dioxide by volume (that is, molar basis). The biogas reacts with 1.2 kmol of oxygen to form carbon dioxide and water. The enthalpy of reaction for methane is $-890,330$ kJ/kmol at 298 K. Calculate the enthalpy change if the reactants are at 298 K and the products are at 1500 K.
>
> The complete reaction is
>
> $$(0.6CH_4 + 0.4CO_2) + 1.2O_2 \rightarrow CO_2 + 1.2H_2O$$

From Table 2.1 the following sensible enthalpies are found:

Temperature (K)	$h_{CH_4}(T)$(kJ/kmol)	$h_{O_2}(T)$(kJ/kmol)	$h_{CO_2}(T)$(kJ/kmol)	$h_{H_2O}(T)$(kJ/kmol)
298	–	8,682	9,364	9,904
1500	–	–	71,078	57,999

Substituting values into Eq. 2.13:

$$[1(71{,}078 - 9{,}364) + 1.2(57{,}999 - 9{,}904)] + 0.6(-890{,}330)$$
$$- [0.6(0) + 0.4(0) + 1.2(0)]$$
$$= -41{,}477$$

Thus, 41,477 kJ is released by the combustion of 1 kmol of biogas under these conditions.

For some well-characterized fuels, such as hydrogen, methane, and ethanol, enthalpies of reaction can be calculated from tabulations of specific enthalpies of formation, $\Delta \bar{h}_f^o$, of chemical compounds from their elements at standard state:

$$\Delta H_R^o = \sum_p n_p \, \Delta \bar{h}_f^o)_p - \sum_r n_r \, \Delta \bar{h}_f^o)_r \qquad (2.14)$$

where n_r and n_p are the stoichiometric coefficients for reactants and products of a chemical reaction. However, many biomass fuels are not well characterized in terms of their chemical constituents because of variations in composition or the presence of complicated chemical compounds. As a result, it is often simpler to perform calorimetric tests on biomass fuels to determine enthalpy of reaction.

Example: Use standard enthalpies of formation to calculate the enthalpy of reaction of liquid ethanol (C_2H_5OH) with oxygen to form carbon dioxide and water vapor.

The stoichiometric reaction is expressed by:

$$C_2H_5OH(\ell) + 3O_2 \rightarrow 2CO_2 + 3H_2O(g)$$

From Table 2.1:

Compound	$C_2H_5OH(\ell)$	O_2	CO_2	$H_2O(g)$
$\Delta \bar{h}_f^o$ (kJ/kmol)	$-277{,}690$	0	$-393{,}520$	$-241{,}820$
Stoichiometric coefficient	1	3	2	3

Substituting into Eq. 2.14 yields:

$$\Delta H_R^o = 2(-393,520) + 3(-241,820) - (-277,690) - 3(0)$$
$$= -1,234,800\text{kJ}$$

2.2.3 Thermodynamic Efficiency

The conversion of chemical energy stored in biomass into more useful forms, such as gaseous and liquid fuels or electrical power, is accompanied by loss of energy to forms that are not easily recovered or utilized. There are many reasons for such losses. Separation processes can inadvertently reject valuable fractions of a feedstock to waste streams. Heat losses can reduce the amount of energy available to energy conversion processes. Entropy production inherent in even ideal processes limits the amount of energy that can be converted into useful forms. Every energy conversion process can be characterized by its thermodynamic efficiency defined as:

$$\eta = \frac{E_{out}}{E_{in}} \tag{2.15}$$

where E_{in} = all forms of energy entering the conversion process
              E_{out} = useful energy leaving the conversion process

These energy flows represent net energy entering and leaving the process. Energy entering the process includes the chemical energy content of the feedstock, thermal energy needed for chemical reactors and distillation units, and power to drive mechanical devices such as tub grinders, pumps, and blowers. The useful energy leaving the process includes chemical energy of gaseous and liquid fuels, electric power, and heat that can be used external to the process, such as district heating of buildings. Obviously, energy must be conserved and the fact that most conversion processes have efficiencies substantially less than unity indicates that some energy leaves the process as waste energy (i.e., unconverted feedstock or waste heat).

Calculation of efficiency is simplified for power plants that convert heat into electrical work. In this case E_{in} is simply the heat input rate, calculated as the enthalpy of reaction (MJ/kg) multiplied by the mass flow rate of fuel (kg/s) while E_{out} is the net power measured in megawatts (MJ/s).

In the electric power industry, the efficiency of converting chemical energy into electric power is frequently expressed as the heat rate, defined as:

$$HR = \frac{\dot{Q}}{P} \text{ (Btu/kW-h)} \tag{2.16}$$

where $\dot{Q}$ is the thermal energy input measured in English units (Btu/h) and P is the net electrical power output measured in SI units (kW). Heat rate is closely related to thermodynamic efficiency except that it is expressed in mixed systems of units. Notice that low heat rates correspond to high thermodynamic efficiencies.

2.2.4 Chemical Equilibrium

A reaction is in chemical equilibrium when the reverse reaction rate balances the forward reaction rate. Two thermodynamic properties important in understanding chemical equilibrium are entropy $\bar{s}(T, p)$ and the Gibbs function $\bar{g}(T, p)$, also known as the Gibbs free energy. These properties, like enthalpy, are tabulated at one atmosphere pressure for various chemical compounds as functions of temperature. Changes in these properties for a chemical reaction can be calculated in a manner similar to calculating changes in enthalpy:

$$\Delta S = \sum_p n_p \bar{s}_p - \sum_r n_r \bar{s}_r \tag{2.17}$$

$$\Delta G = \sum_p n_p \bar{g}_p - \sum_r n_r \bar{g}_r \tag{2.18}$$

These two properties are related by the following equation:

$$\Delta G = \Delta H - T\Delta S \tag{2.19}$$

where the temperature T is evaluated in Kelvin.

The Gibbs function is particularly useful in chemical thermodynamics. First, the change in Gibbs function represents the maximum work that could be produced from a chemically reacting system. As is evident from Eq. 2.19, the Gibbs function represents some fraction of the chemical enthalpy associated with a chemical reaction, an important consideration in the discussion of fuel cells in Chapter 5. Second, the Gibbs function is useful in calculating the equilibrium composition of a chemically reacting system. From the Second Law of Thermodynamics, it can be shown that chemical equilibrium for a constant pressure, constant temperature process corresponds to a minimum in the Gibbs function. For a reaction involving ideal gases, this relationship is given by:

$$K_p(T) = \frac{\prod_p \left(\dfrac{p_p}{p_{\text{ref}}} \right)^{\nu_p}}{\prod_r \left(\dfrac{p_r}{p_{\text{ref}}} \right)^{\nu_r}} \tag{2.20}$$

where p_p and ν_p are the partial pressure and stoichiometric coefficient, respectively, of the pth chemical product, and p_r and ν_r are the partial pressure and stoichiometric coefficient, respectively, of the rth chemical reactant; p_{ref} is the reference pressure (usually one atmosphere); and $K_p(T)$ is the equilibrium constant in terms of partial pressures as a function of temperature T. The symbol Π is mathematical shorthand that indicates the calculation of a product among the partial pressure terms of the chemical products or reactants. The equilibrium constant is defined by the expression:

$$\ell_n K(T) = -\frac{\Delta G^o}{\overline{R}T} \qquad (2.21)$$

where the superscript on the Gibbs function indicates that it is evaluated at the reference pressure and the indicated temperature.

2.3 Organic Chemistry

The original distinction between inorganic and organic compounds was their source in nature. Inorganic compounds were derived from mineral sources whereas organic compounds were obtained from plants or animals. Advances in chemical synthesis since the eighteenth century have made this distinction obsolete as the vast majority of organic chemicals commercially produced today are made from petroleum. The common feature of organic compounds is a skeleton of carbon atoms that includes lesser amounts of other atoms, especially hydrogen, oxygen, and nitrogen, but also sulfur, phosphorus, and halides.

The high chemical valence of carbon allows for complex structures and large numbers of organic compounds. These include compounds consisting of chains of carbon atoms, referred to as acyclic or aliphatic compounds, and compounds containing rings of carbon atoms, known as carbocyclic or simply cyclic compounds. Some of these rings contain at least one atom that is not carbon (known as heteroatoms). These compounds are called heterocyclic compounds. Carbocyclic compounds are further classified as either aromatic compounds, in which electrons are shared among atoms to produce a particularly stable ring, or alicyclic compounds, which includes all non-aromatic cyclic compounds.

A variety of reactions can occur among organic compounds. *Addition* reactions occur when two reactants combine to give a single product. *Elimination* reactions involve the splitting of a single compound into two compounds. Most elimination reactions form a product with a double bond containing the majority of the atoms found in the reactant. *Substitution* reactions involve replacement of one atom or group of atoms by a second atom or group of atoms. *Hydrolysis* reactions involve the action of water in splitting a large reactant molecule into two smaller product molecules. One product molecule is bonded to the hydrogen atom from the water while the other product molecule is bonded to the hydroxyl group derived from the water. *Condensation* reactions involve two reactants combining to form one larger product with the simultaneous formation of a second, smaller product such as water. *Rearrangement* reactions result from the reorganization of bonds within a single reactant to give an isomeric product.

2.3.1 Structural Formulas and Chemical Nomenclature

The molecular formula of a compound indicates its atomic composition. For example, the molecular formula for pentane is C_5H_{12}. Structural formulas show the arrangement of atoms and bonds. The number of lines between atoms in structural formulas indicates the number of bonds between them. For example, a carbon-hydrogen bond is represented by C–H, a carbon-carbon double bond is represented by C=C, and a carbon-carbon triple bond is written C≡C. Condensed structural formulas show only specific bonds; other bonds are left out but implied. The degree of condensation of structural formulas is somewhat arbitrary. Commonly, C–H bonds are not shown because they can only form single bonds. Additional condensation of structural formulas can be achieved by omitting C–C bonds.

C_5H_{12}

$$H-\overset{\overset{\displaystyle H}{|}}{\underset{\underset{\displaystyle H}{|}}{C}}-\overset{\overset{\displaystyle H}{|}}{\underset{\underset{\displaystyle H}{|}}{C}}-\overset{\overset{\displaystyle H}{|}}{\underset{\underset{\displaystyle H}{|}}{C}}-\overset{\overset{\displaystyle H}{|}}{\underset{\underset{\displaystyle H}{|}}{C}}-\overset{\overset{\displaystyle H}{|}}{\underset{\underset{\displaystyle H}{|}}{C}}-H$$

$CH_3-CH_2-CH_2-CH_2-CH_3$ $CH_3(CH_2)_3CH_3$

molecular formula   structural formula   condensed structural formula   additional condensation of structural formula

Bond-line structures are an extreme shorthand for representing molecules. Carbon atoms are omitted but carbon-carbon bonds are illustrated in a zigzag arrangement. Carbon-hydrogen bonds are omitted while bonds of carbon atoms to other atoms or molecular groups are shown explicitly.

$CH_2CH(CH_2)_2COCH_3$

condensed structural formula   bond-line structure

Sometimes it will be important to distinguish the three-dimensional structure of a molecule. In this case, perspective structural formulas are employed, with wedges indicating bonds that project out of the plane of the drawing. Solid wedges indicate bonds projecting above the plane of the drawing while wedges shaded with parallel lines indicate bonds projecting below the plane of the drawing.

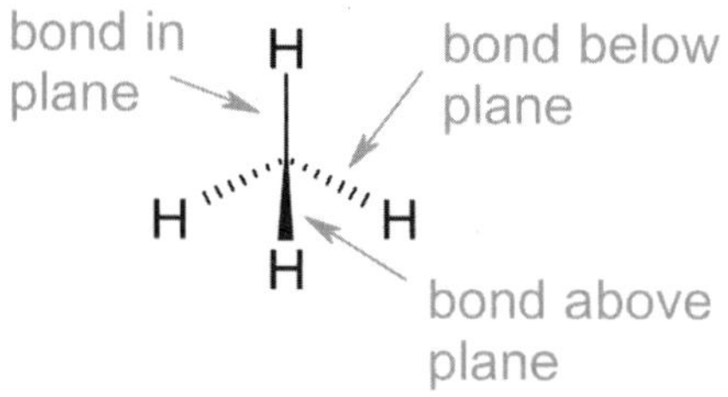

bond in plane → H (bond below plane)

H ''''''|'''''' H

H

bond above plane

perspective structural
formula for methane

Organic compounds are named according to a system devised by the International Union of Pure and Applied Chemistry (IUPAC). The IUPAC name consists of three parts: prefix, parent, and suffix. The parent identifies the main carbon chain. The suffix identifies most of the functional groups present in the molecule. The prefix specifies the location of functional groups identified in the suffix as well as identifies some other atoms or groups of atoms attached to the main carbon chain. The nomenclature of organic compounds is further elaborated in the subsequent discussions on various functional groups. Many compounds continue to be known by their common names either because of long usage or to avoid the unwieldy nomenclature of some compounds derived from biological sources.

Compounds that have the same molecular formula but different structures are isomers. Isomers that differ in their carbon skeleton are called skeletal isomers. For example, butane and isobutane both have the molecular formula of C_4H_{10} but butane is a straight-chain hydrocarbon whereas isobutane includes a short side-chain.

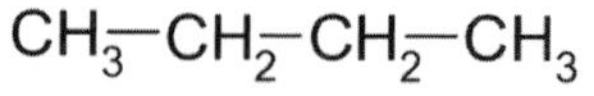

$CH_3-CH_2-CH_2-CH_3$ $CH_3-\overset{\overset{\displaystyle CH_3}{|}}{CH}-CH_3$

butane 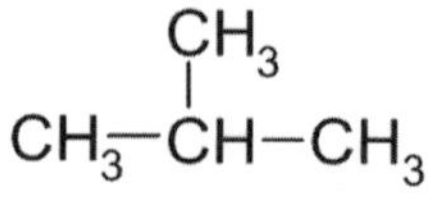   isobutane

Isomers that have different functional groups are called functional group isomers. For example the molecular formula C_2H_6O represents both ether, which has a functional group in the form of divalent oxygen, and an alcohol, which has a functional group in the form of monovalent hydroxyl group (OH).

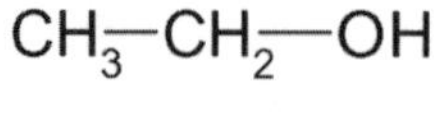

CH_3-CH_2-OH CH_3-O-CH_3

ethanol 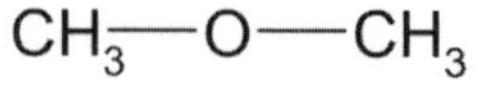   dimethyl ether

Positional isomers have both the same molecular formulas and functional groups. They differ only in the location of the functional groups in the carbon chains. For example, 1-propanol and 2-propanol are alcohols of molecular formula C_3H_7OH but the former has its single hydroxyl group located at the end of the carbon chain whereas the latter has its hydroxyl group at the middle of the chain.

$$CH_3-CH_2-CH_2-OH \qquad\qquad CH_3-\overset{\overset{\displaystyle OH}{|}}{CH}-CH_3$$

1-propanol   2-propanol

Organic compounds are conveniently classified according to functional groups: the molecular structures that give the compounds distinctive chemical and physical properties. Those functional groups of particular interest to the chemistry of biobased products are summarized in Table 2.2. The rest of this section describes the chemistry of these functional groups.

2.3.2 Alkanes

The simplest organic compounds consist only of carbon and hydrogen atoms and are known as *hydrocarbons*. Hydrocarbons containing only carbon-carbon single bonds are classified as *alkanes* and conform to the general chemical formula of C_nH_{2n+2}. Alkanes are also called *paraffins*. A hydrocarbon that contains only carbon-carbon single bonds is said to be *saturated*.

Alkanes with a continuous chain of carbon atoms are normal alkanes. The names and molecular formulas for normal alkanes containing up to 10 carbon atoms are listed in Table 2.3. Notice that all end in the suffix -ane. The first four names are based on common names while those with more than four carbon atoms are derived from Greek numbers that indicate the number of carbon atoms. Familiar alkanes include the fuels methane (CH_4), propane (C_3H_8), and butane (C_4H_{10}). Gasoline is a mixture of higher boiling point alkanes.

A normal alkane molecule minus one hydrogen atom is known as an alkyl group; for example, CH_3- is the methyl group. As shown in Table 2.3, the names of the alkyl groups conform to the names of the corresponding alkanes. The short hand for an alkyl group is $R-$ (that is, "remainder of molecule"). The number of alkyl groups attached to hydrocarbons is a convenient way to classify them.

Branched alkanes result when carbon atoms in the chain bond to other carbon atoms, forming side chains. These still have the general formula C_nH_{2n+2} and thus represent isomers of the normal alkanes. They are named such that the longest continuous chain is the parent part of the IUPAC name. The name and position of the

Table 2.2. Classification of organic compounds based on functional groups

Class	General Formula	Functional Group	Suffix To Chemical Name
Alkane	RH	$C-H$ $C-C$	-ane
Alkene	RCH=CH$_2$ RCH=CHR R$_2$C=CHR R$_2$C=CR$_2$	$\diagdown C = C \diagup$	-ene
Alkyne	RC≡CH RC≡CR	$-C≡C-$	-yne
Aromatic hydrocarbon	ArH	(benzene ring)	-
Alcohol or phenol	ROH	$-O-H$	-ol
Ether	ROR	$-C-O-C-$	-
Aldehyde	RCHO	$-\overset{O}{\overset{\|\|}{C}}-H$	-al
Ketone	RCOR	$-\overset{O}{\overset{\|\|}{C}}-$	-one
Carboxylic acid	RCOOH	$-\overset{O}{\overset{\|\|}{C}}-O-H$	-oic acid
Ester	RCOOR	$-\overset{O}{\overset{\|\|}{C}}-O-C-$	-oate
Amine	RNH$_2$	$-\overset{H}{\overset{\|}{N}}-H$	-amine
Amide	RCONH$_2$	$-\overset{O}{\overset{\|\|}{C}}-\overset{H}{\overset{\|}{N}}-H$	-amide

Table 2.3. Names of alkanes and alkyl groups

Number of Carbon Atoms	Alkane		Alkyl Group	
	Name	Molecular Formula	Name	Molecular Formula
1	Methane	CH$_4$	Methyl	CH$_3$–
2	Ethane	C$_2$H$_6$	Ethyl	C$_2$H$_5$–
3	Propane	C$_3$H$_8$	Propyl	C$_3$H$_7$–
4	Butane	C$_4$H$_{10}$	Butyl	C$_4$H$_9$–
5	Pentane	C$_5$H$_{12}$	Pentyl	C$_5$H$_{11}$–
6	Hexane	C$_6$H$_{14}$	Hexyl	C$_6$H$_{13}$–
7	Heptane	C$_7$H$_{16}$	Heptyl	C$_7$H$_{15}$–
8	Octane	C$_8$H$_{18}$	Octyl	C$_8$H$_{17}$–
9	Nontane	C$_9$H$_{20}$	Nontyl	C$_9$H$_{19}$–
10	Decane	C$_{10}$H$_{22}$	Dectyl	C$_{10}$H$_{21}$–

alkyl groups making up the side chains are included in the suffix part of the IUPAC name. Positions of alkyl groups are specified by numbering carbon atoms along the main carbon chain, with the lowest numbered carbon atom appearing at the end of the chain closest to the first branch chain. The prefix specifies all alkyl groups attached to the main chain, with the prefixes listed alphabetically rather than by location on the main chain. Two or more groups of the same name are indicated by the prefix di-, tri-, tetra-, etc. For example, an alkane consisting of five carbon atoms in the main chain with a methyl group attached to the carbon atom second to the end of the main chain is called 2-methylpentane. Replacement of an ethyl group for one of the hydrogen atoms attached to the third carbon of 2-methylpentane is called 3-ethyl-2-methylpentane. Notice that if the ethyl group had been attached to the fourth carbon atom it would not be called 4-ethyl-2-methylpentane since the ethyl group now becomes part of the longest chain in the molecular structure. Instead, it should be considered as consisting of a six-carbon main chain with two methyl groups attached to the second and fourth carbon atoms, which is called 2,4-dimethylhexane.

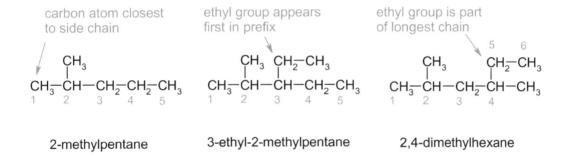

Closely related to the alkanes are the *cycloalkanes*, which contain carbon-carbon single bonds in a ring structure and conform to the general chemical formula of C_nH_{2n}. The simplest cycloalkanes are cyclopropane (C_3H_6) in a triangular ring, cyclobutane in a rectangular ring (C_4H_8), cyclopentane (C_5H_{10}) in a pentagonal ring, and cyclohexane (C_6H_{12}) in a hexagonal ring.

Since C–H bonds are non-polar, alkanes are non-polar, with implications for the physical properties of these hydrocarbons. Alkanes are not soluble in water, a highly polar compound (following the rule-of-thumb that "like dissolves like"). The absence of polarity also means that alkane molecules do not strongly interact; thus, boiling points of alkanes are relatively low compared to other organic compounds, increasing with molecular weight. The C–C and C–H bonds of alkanes are not very reactive with the exception of oxidation. The high heats of oxidation make alkanes attractive as fuels.

2.3.3 Alkenes and Alkynes

Hydrocarbons with carbon-carbon double bonds are called *alkenes* and conform to the general chemical formula of C_nH_{2n}. The IUPAC nomenclature follows that of the alkanes (see Table 2.3) except the suffix -ene is used instead of -ane. Like alkanes, alkenes can have branched chains. The longest continuous chain containing the double bond is the parent. Carbon atoms are numbered consecutively from the end nearest to the double bond. Isomers are distinguished by including the number of the first carbon atom in the double bond in the prefix to the parent name. The prefix also specifies the type and location of all alkyl groups attached to the main chain, with the prefixes listed alphabetically rather than by location on the main chain. Two or more groups of the same name are indicated by the prefix di-, tri-, tetra-, etc.

2-pentene   2-methyl-2-pentene   2,3-dimethyl-2-pentene

Common names are often employed in the chemical industry. For example, the simplest alkene, ethene (C_2H_4), is commonly known as ethylene. Ranked first among organic chemicals in annual production in the United States, ethene is produced from natural gas liquids.

Alkenes are widely used in the production of polymers, a high molecular weight compound created by the repetitive reaction of low molecular weight molecules, known as monomers, to form long carbon chains. Polymerization of alkenes occurs by addition reactions: carbon chains are lengthened by addition of monomers without the formation of other products. For example, ethene, at elevated pressures and temperatures and in the presence of a trace amount of oxidant, yields a poly-

meric chain of molecular mass of 50,000 to 300,000 known as polyethene (familiarly known as polyethylene), a low density plastic used in the manufacture of plastic bags and squeeze bottles.

$$n CH_2{=}CH_2 \longrightarrow {-}[CH_2{-}CH_2]_n{-}$$

ethene   polyethene chain

Other polymers from alkenes include polypropene, commonly called
polypropylene and spun into fibers for carpets and ropes; polychloroethene, also
known as polyvinyl chloride, widely used in making both rigid plastic products,
such as pipe and floor tiles, as well as flexible plastic products, such as shower cur-
tains and garden hoses; polydichloroethene, used in plastic wrap; polytetrafluo-
roethene, the basis of Teflon; polyacrylonitrile, used in synthetic fibers;
polystyrene, used in Styrofoam; and polymethyl methacrylate, used in Plexiglass.

If more than one double bond appears in a molecule, the location of each double bond is indicated in the prefix and the suffix becomes -diene, -triene, -tetraene
as appropriate to indicate the number of double bonds. Alkadienes, containing two
double bonds, are commonly called dienes. If the two double bonds of a diene are
separated by one single bond, the resulting compound, known as a conjugated
diene, has chemical properties distinct from other alkenes (if separated by more
than one single bond, the double bonds are unconjugated dienes and have properties similar to other alkenes). An important example of a conjugated diene is 2-
methyl-1,3-butadiene, commonly known as isoprene.

structural formula   bond-line

2-methyl-1,3-butadiene (isoprene)

Isoprene is the basis for many important natural chemicals. Both terminal carbon atoms can bond with terminal carbon atoms of other isoprenes to form a variety of cyclic and acyclic compounds with different degrees of saturation and

various functional groups attached. Terpenes, compounds of two or more isoprene units joined together with the generic formula $(C_5H_8)_n$, have distinctive odors and flavors. They are responsible for the fragrant odors of pine trees and the bright colors of tomatoes and carrots. Terpenes are classified by the number of isoprene units they contain: monoterpenes have two isoprene units, sesquiterpenes have three isoprene units, and diterpenes, triterpenes, and tetraterpenes contain 4, 6, and 8 isoprene units, respectively. Examples include farnesol, an acyclic sesquiterpene first isolated from roses and citronella, and limonene, a cyclic monoterpene with a distinctive lemon odor.

dashed lines separate the isoprene units

farnesol   limonene

Hydrocarbons with carbon-carbon triple bonds are called *alkynes* and conform to the general chemical formula C_nH_{2n-2}. The nomenclature follows that of the alkanes (see Table 2.3) except the suffix -yne is used instead of -ane, although common names are often employed in the chemical industry. For example, the simplest alkyne, ethyne (C_2H_2), is commonly known as acetylene, the fuel for many high-temperature cutting torches. Compounds with multiple triple bonds are diynes, triynes, and so on. Compounds with both double and triple bonds are called enynes; numbering of these compounds starts from the end nearest the first multiple bond.

The presence of double and triple bonds in alkenes and alkynes, respectively, makes them unsaturated hydrocarbons. These unsaturated compounds are often classified according to the number of alkyl groups (R–) attached to the double or triple bond unit, which is known as the "degree of substitution." As an example, various degrees of substitution are illustrated in Table 2.4 for an alkene and an alkyne.

Both alkenes and alkynes are non-polar compounds with physical properties similar to those for alkanes. However, the double and triple bond components in alkenes and alkynes are weaker than the single bond in alkanes, making them more chemically reactive than alkanes. Unsaturated compounds are easily oxidized without destroying the carbon chain. For example, potassium permanganate readily

Table 2.4. Examples of substitutions of alkyl groups in alkenes and alkynes

Substitution	Ethene	Ethyne
Unsubstituted	H₂C=CH₂ (H,H on each carbon)	H—C≡C—H
Mono-substituted	R,H C=C H,H	R—C≡C—H
Di-substituted	R,R C=C H,H	R—C≡C—R
Di-substituted	R,H C=C H,R	
Tri-substituted	R,R C=C H,R	
Tetra-substituted	R,R C=C R,R	

oxidizes ethene to 1,2-ethanediol, commonly known as ethylene glycol, familiar as automotive anti-freeze.

$$CH_2 = CH_2 \quad \xrightarrow[\text{H}_2\text{O}]{\text{KMnO}_4} \quad \begin{array}{l} CH_2-OH \\ | \\ CH_2-OH \end{array}$$

ethene   1,2-ethanediol

An important reduction reaction, known as *hydrogenation*, is the conversion of a C=C bond in an alkene or the C≡C bond in an alkyne to a C–C bond by the addition of hydrogen in the presence of a catalyst.

$$\begin{array}{c} H \\ H \end{array} C=C \begin{array}{c} H \\ H \end{array} \; + \; H_2 \quad \longrightarrow \quad H-\overset{H}{\underset{H}{C}}-\overset{H}{\underset{H}{C}}-H$$

ethene   ethane

2.3.4 Aromatic Compounds

Among the most important cyclic hydrocarbon structures is the *benzene ring*, consisting of six carbon atoms in an extremely stable hexagonal structure.

benzene: structural formula   benzene: bond-line structure

This stability is the result of the sharing of electrons (*delocalization*) among all the atoms of the ring, a condition that can be predicted by the Hückel rule. Compounds that satisfy the Hückel rule are known as aromatic compounds based on the fact that many of these compounds have distinctive fragrances, such as vanilla and oil of wintergreen. Compounds based on the six-carbon ring of benzene are not the only cyclic compounds classified as aromatic, but they are among the most important in organic chemistry.

Substitution of various chains or ring structures for the hydrogen atoms yields a tremendous variety of chemical compounds with a range of useful properties. For example, substitution with a hydroxyl group yields phenol, a crystalline acidic compound used in the manufacture of phenolic resins and readily produced by the pyrolysis of biomass. Substitution with an amine group yields aniline, the basis of the synthetic dye industry in the nineteenth century. Substitution with a carboxylic group yields benzoic acid, a white crystalline acid found naturally in cranberries and used especially as a preservative of foods.

phenol   aniline   benzoic acid

If one or more of the carbon atoms in the ring of an aromatic compound are substituted by another atom, usually nitrogen or oxygen in naturally occurring compounds, the resulting compound is said to be a heterocyclic aromatic compound. Examples include pyridine, a pungent, water-soluble flammable liquid that is the parent of many naturally occurring organic compounds, and furan, a flammable liquid that is obtained from pine wood:

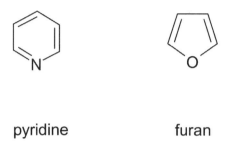

pyridine   furan

Aromatic compounds consisting of two or more rings "fused" together are known as polycyclic aromatic hydrocarbons. These compounds are planar, that is, all atoms in the rings and those atoms directly attached to the rings are in a plane. A prominent example is naphthalene, often used in the manufacture of mothballs:

naphthalene

An aromatic ring attached to a larger parent structure is an aryl group, often symbolized by Ar–, just as the alkyl group is symbolized by R–. Examples of aryl groups derived from benzene are the phenyl group and the benzyl group:

phenyl group   benzyl group

2.3.5 Alcohols and Phenols

Hydrocarbons, compounds of only carbon and hydrogen, make up only a small fraction of plant material. Most plant material incorporates large amounts of oxygen and, thus, functional groups containing oxygen play important roles in the chemistry of biobased products. Organic compounds that have functional groups containing oxygen include alcohols, phenols, ethers, aldehydes, ketones, acids, esters, and amides.

The *hydroxyl group*, $-OH$, characterizes alcohol, and members of this group are named according to the longest carbon chain containing the hydroxyl group. The parent name is obtained by adding -ol to the name of the alkyl group attached to the hydroxyl group. The location of the hydroxyl group in the parent chain is specified by adding, as a prefix, a number corresponding to the carbon atom to which the hydroxyl group is attached. The lowest numbered carbon atom is the one at the end of the carbon chain that is located closest to the hydroxyl group. For example, an alcohol formed from the attachment of a hydroxyl group to the end of the butyl group is known as 1-butanol whereas a hydroxyl group attached to the second of the four carbon atoms in the butyl group is called 2-butanol.

1-butanol   2-butanol

Some alcohols, known as *polyols*, contain two or more hydroxyl groups. They are named by retaining the -e in the name of the parent alkane and adding -diol, -triol, and so forth to indicate the number of hydroxyl groups attached, although many polyols have common names. For example, 1,2-ethanediol consists of two hydroxyl groups attached to an ethane background. Commonly known as ethylene glycol, it is used both as anti-freeze and an important chemical in the manufacture of the synthetic fiber Dacron® and Mylar® film. An example of a triol is 1,2,3-propanetriol, commonly known as glycerol, which consists of three hydroxyl groups attached to a propane chain, and is the backbone of fats and oils.

1,2-ethanediol   1,2,3-propanetriol

Alcohols are also formed by attaching hydroxyl groups to ring structures and to carbon chains containing double or triple bonds. The carbon atoms on ring structures are numbered starting with the carbon atom attached to the hydroxyl group and continuing around the ring in the direction that gives the lowest numbers to atoms attached to substituents. Alcohols containing double and triple bonds are named by including a prefix number that labels the position of the double bond and a suffix that combines the position of the hydroxyl group along with the designation -ol.

$$H_2C=CH-CH_2-CH_2-OH$$

3-buten-1-ol   4,4-Dimethyl-cyclohexanol

The hydroxyl group is responsible for several prominent physical properties of alcohols. The hydroxyl group serves as both a hydrogen bond donor and a hydrogen bond acceptor, resulting in large intermolecular forces between alcohol molecules. The large energy required to separate these bonds is responsible for high boiling points of alcohols compared to alkanes of similar molecular weight. Like water, lower molecular weight alcohols are highly polar, resulting in high solubility in water and in other polar organic compounds.

Alcohols can serve as both proton donors and proton acceptors. When an alcohol loses a proton, a conjugate base called an alkoxide ion, R-O$^-$, is produced. Alcohols are very weak acids. When an alcohol accepts a proton, a conjugate acid called an oxonium ion, R-OH$_2$$^+$, is produced. Alcohols are also very weak bases, forming only under the action of a very strong acid.

Alcohols are classified as primary, secondary, or tertiary alcohols depending upon how many carbon atoms are attached to the carbon atom that is bonded to the hydroxyl group:

primary alcohol   secondary alcohol   tertiary alcohol

Removal of a water molecule from an alcohol is a dehydration reaction, which is an example of an elimination reaction. This reaction requires an acid catalyst, such as sulfuric acid or phosphoric acid, and is illustrated by the formation of ethylene from ethanol.

Primary alcohols, which have the general formula of RCH_2OH, can be oxidized to *aldehydes*, a class of organic compounds with the general formula of RCHO; the reaction is accompanied by the removal of two hydrogen atoms. The resulting aldehydes can be further oxidized to carboxylic acids, a class of organic compounds with the general formula of RCOOH; the oxidized molecule gains an oxygen atom. Secondary alcohols, which have the general formula of R_2CHOH, are oxidized to *ketones*, a class of organic compounds with the general formula of RCOR, which cannot be further oxidized because there is no hydrogen atom on the oxygen-bearing carbon atom of the ketone. Tertiary alcohols, which have the general formula R_3COH, cannot be readily oxidized because the carbon atom bearing the hydroxyl group has no hydrogen atom.

Alcohols can be synthesized from alkenes by hydration (addition of water). Conversely, alkenes can be dehydrated (removal of water) to form alcohols. For example, the equilibrium expression for hydration of ethene and dehydration of ethanol is given by:

ethene   ethanol

The direction of the reaction is controlled by Le Chatelier's principle. Thus, excess water will drive the reaction toward alcohol formation whereas a deficiency in water will drive the reaction toward alkene formation. This latter condition is achieved by reaction in concentrated sulfuric acid, where water concentration is very low.

The attachment of a hydroxyl group to benzene forms an important class of compounds known as *phenols*. Both alcohols and phenols can be considered organic analogs of water and undergo reactions that are similar to water, such as reaction with alkali metals to form hydrogen gas. However, the delocalization of electrons in the benzene ring makes the C–O bond in phenols much stronger than the C–O bond in alcohols, resulting in important chemical differences between these two classes of compounds. In alcohols, the hydroxyl group is readily displaced whereas the strong C–O bond in phenol prevents this from occurring except under extreme conditions.

Examples of phenolic compounds include phenol, with one hydroxyl group attached to the benzene ring; pyrocatechol, containing two adjacent hydroxyl groups and originally obtained from pyrolysis of biomass; and urushiol, the oily irritant in poison ivy, which is a mixture of pyrocatechol derivatives with saturated and unsaturated side chains of 15 or 17 carbon atoms. Phenol is used to manufacture epoxy resins and polymers such as Bakelite and nylon. It is also used in the manufacture of dyes, herbicides, and disinfectants.

| phenol | pyrocatechol | generic urushiol |

2.3.6 Ethers

Ethers contain two groups, either alkyl or aryl groups, bonded to an oxygen atom. The smaller alkyl group and the oxygen atom constitute an alkoxy group (R-O). Similarly, an aryl group and the oxygen atom constitute a phenoxy group (Ar-O). The alkoxy and phenoxy groups are considered substituents on the larger parent alkyl or aryl group. These groups can form either symmetrical ethers or unsymmetrical ethers, depending upon whether the same or different alkyl or aryl groups are attached to the two oxygen bonds. Ethers are named by listing the alkyl or aryl groups in alphabetical order and appending the name ether.

| diethyl ether | ethyl methyl ether | diphenyl ether |
| (symmetrical ether) | (asymmetrical ether) | (symmetrial ether) |

The two C–O bonds result in substantial dipole moments for ethers. They are more polar than alkanes but less polar than alcohols. Thus, they are more soluble in water than alkanes of similar molecular weight. This functional group has no hydrogen bond donors; thus, boiling points are substantially less than for alcohols of comparable molecular weight and similar to alkanes of comparable molecular weight. Ethers are stable compounds that do not react with most common reagents. Ethers have found applications as solvents and anesthetics.

2.3.7 Aldehydes and Ketones (Carbonyl Compounds)

Aldehydes and ketones are characterized by the carbonyl functional group, which consists of an oxygen atom double-bonded to carbon (C=O). The carbon atom of the carbonyl group is called the carbonyl carbon atom, and the oxygen atom is called the carbonyl oxygen atom. The carbonyl group is prevalent in compounds isolated from plants. Aldehydes and ketones, collectively known as carbonyl compounds, often have pleasant odors and are responsible for the fragrant smell of flowers. At one time plants were the sole source of aldehydes and ketones used in such products as perfumes.

An *aldehyde* is a carbonyl group with the carbonyl carbon atom bonded to a hydrogen atom and either another hydrogen atom or an alkyl or aryl group. The aldehydes are named by addition of the suffix -al to the root alkyl name corresponding to the longest continuous carbon chain containing the carbonyl group. For example, methanal is a carbonyl group (C=O) bonded to two hydrogen atoms. Methanal is more commonly known as formaldehyde, a preservative for biological specimens. Ethanal, containing two carbon atoms, consists of a carbonyl group bonded to one hydrogen atom and one methyl group (CH$_3$). Ethanal is more commonly known as acetaldehyde, a compound found in a variety of fruits and vegetables.

$$\overset{\displaystyle O}{\underset{\displaystyle H-C-H}{\|}} \qquad \overset{\displaystyle O}{\underset{\displaystyle CH_3-C-H}{\|}} \qquad \overset{\displaystyle O}{\underset{\displaystyle R-C-H}{\|}}$$

methanal   ethanal   generic aldehyde

The alkyl group attached to the carbonyl group may also have side chains, which must be indicated in the name. This is done by specifying the location and type of alkyl group as a prefix to the parent name. Carbon atoms in the parent chain are numbered starting with the carbonyl carbon atom. The presence of carbonyl side chains (C=O) is designated with the prefix oxo-.

4-hydroxyl-2-methyl-pentanal   2-methyl-4-oxo-pentanal

An example of an aldehyde incorporating an aryl group is benzaldehyde, consisting of a carbonyl group bonded to a hydrogen atom and a phenyl group (that is, a benzene ring). Benzaldehyde is known as oil of almonds and used in flavorings and perfumes.

benzaldehyde

When the carbonyl carbon atom is bonded to two other carbon atoms, the compound is known as a *ketone*. These other carbon atoms can be part of either an alkyl or aryl group. The ketones are named by addition of the suffix -one to the root alkyl name corresponding to the longest continuous carbon chain containing the carbonyl group. Replacing the hydrogen bond in ethanal, for example, with another methyl group yields the solvent propanone (commonly known as acetone) while replacing the hydrogen bond in benzaldehyde with another phenyl group yields benzophenone, a component in the manufacture of perfumes and sunscreens.

propanone   benzophenone   generic ketone

For more complicated ketones, the location of the carbonyl carbon atom and the position and type of side chains must be specified. The carbon atoms are numbered from the end of the parent chain closest to the carbonyl carbon atom. Cyclic ketones are called cycloalkanones. The carbonyl carbon atom is designated as the first carbon atom in the ring.

3-methyl-2-butanone   3-methylcyclohexanone

The carbonyl group is polar; thus, carbonyl compounds have boiling points that are higher than alkanes of similar molecular weight. The carbonyl group cannot act as a hydrogen donor but the carbonyl oxygen atom can serve as a hydrogen acceptor, allowing aldehydes and ketones to form hydrogen bonds with water. Accordingly, lower molecular weight compounds containing the carbonyl group are miscible in water. For the same reason, alcohols and carboxylic acids are soluble in these compounds. Several ketones are excellent solvents for polar organic compounds.

Aldehydes can be synthesized by the controlled oxidation of primary alcohols while ketones can be prepared by oxidation of secondary alcohols. Carbonyl compounds can participate in a variety of chemical reactions including oxidation to carboxylic acids, reduction to alcohols, and condensation to give a group of compounds known as aldols, which contain both aldehyde and alcohol functional groups. They also participate in acid-catalyzed addition reactions with alcohol to form compounds very important to carbohydrate chemistry.

In the addition reaction of alcohol with a carbonyl compound, the hydrogen atom from the hydroxyl group of the alcohol adds to the carbonyl oxygen atom whereas the remaining –OR' fragment from the alcohol, known as the alkoxy group, adds to the carbonyl carbon atom. The product is called *hemiacetal* if the carbonyl compound in the reactants is aldehyde.

R'—O—H + (H, C=O, R aldehyde) → (R'—O, H—C—O—H, R hemiacetal)

alcohol   aldehyde   hemiacetal

The product is called *hemiketal* if the carbonyl compound in the reactants is ketone.

R'—O—H + (R, C=O, R ketone) → (R'—O, R—C—O—H, R hemiketal)

alcohol   ketone   hemiketal

The equilibrium constant for the formation of either hemiacetal or hemiketal is less than one; thus, these compounds are usually unstable. However, if the hydroxyl and carbonyl groups are part of the same molecule, a stable cyclic product forms.

hydroxyl and carbonyl groups
in same organic compound

stable cyclic hemiacetal
or hemiketal product

Carbohydrates contain both hydroxyl and carbonyl groups and commonly form such cyclic hemiacetals or hemiketals as will be discussed in a later section.

Notice that the resulting hemiacetal or hemiketal contains a new hydroxy group. In an acidic solution, this hydroxyl group can be replaced by another alkoxy group derived from alcohol to form acetal and ketal, respectively.

alcohol   hemiacetal   acetal

alcohol   hemiketal   ketal

The position of the equilibrium can be shifted toward formation of acetal or ketal by adding alcohol or removing water from the reaction. The reverse reactions are acid-catalyzed hydrolysis of acetal or ketal to form carbonyl compounds and alcohol. Cyclic hemiacetyls and cyclic hemiketals react with alcohols to form cyclic acetals and cyclic ketals, respectively.

2.3.8 Carboxylic Acids

The *carboxyl* group, a combination of carbonyl group and hydroxyl group, characterizes carboxylic acids.

$$H_3C-\overset{\overset{\displaystyle O}{\|}}{C}-OH$$

carboxyl group

The carbonyl group and hydroxyl group modify the behavior of each other; thus, the behavior of carboxylic acids differs from that of carbonyl compounds and alcohols.

Carboxylic acids are named by replacing the -e in the root alkyl name corresponding to the longest continuous carbon chain containing the carboxyl group. For example, attached to a hydrogen atom, it forms methanoic acid, commonly known as formic acid, a pungent liquid secreted by ants when disturbed; attached to the methyl group it forms ethanoic acid, commonly known as acetic acid, the main ingredient of vinegar and important in the manufacture of polymers; attached to the propyl group it forms butanoic acid, commonly known as butyric acid, responsible for the unpleasant odor of rancid butter and rotting hay.

$$H-\overset{\overset{\displaystyle O}{\|}}{C}-OH \qquad CH_3-\overset{\overset{\displaystyle O}{\|}}{C}-OH \qquad CH_3-CH_2-CH_2-\overset{\overset{\displaystyle O}{\|}}{C}-OH$$

methanoic acid   ethanoic acid   butanoic acid

Carboxylic acids can have additional functional groups attached to the carbon chain; these are named by adding prefixes that indicate the position of the functional group. The carbonyl carbon atom at the end of the parent chain is designated as the first carbon atom. Examples include 2-hydroxypropanoic acid, commonly known as lactic acid, important in the manufacture of biodegradable plastics, and 4-oxopentanoic acid, commonly known as levulinic acid, which can be used as a fuel additive or in the production of synthetic resins and plastics.

2-hydroxypropanoic acid   4-oxopentanoic acid

A special class of carboxylic acids, known as *fatty acids*, consists of long carbon chains with even numbers of carbon atoms. Fatty acids differ primarily in chain length and degree of saturation. Chains are generally linear and most commonly are 16 or 18 carbons long. They may be saturated, like stearic acid; monounsaturated (containing one carbon-carbon double bond), like oleic acid; or polyunsaturated (containing several carbon-carbon double bonds), like linoleic acid. These are important components of fats and oils.

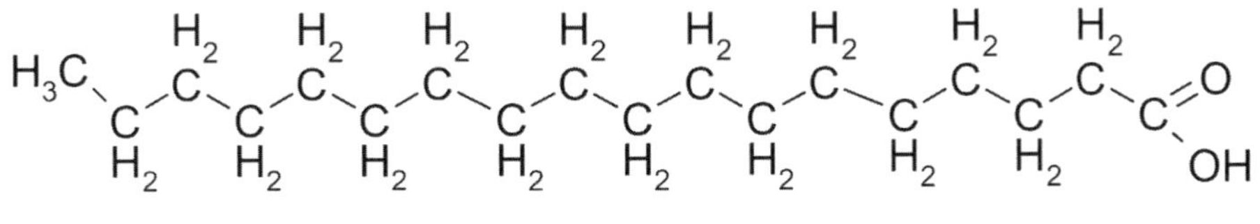

stearic acid

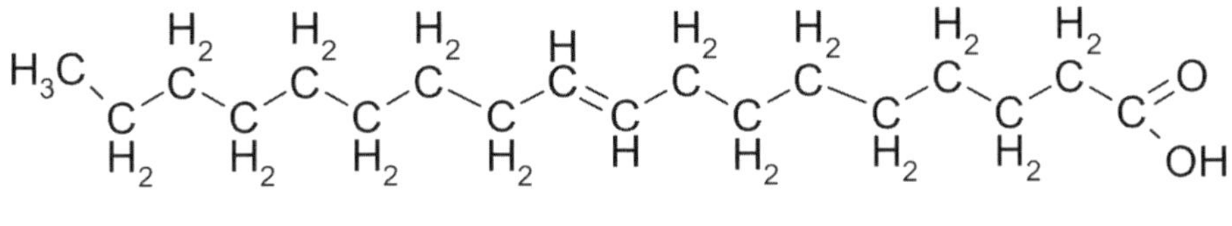

oleic acid

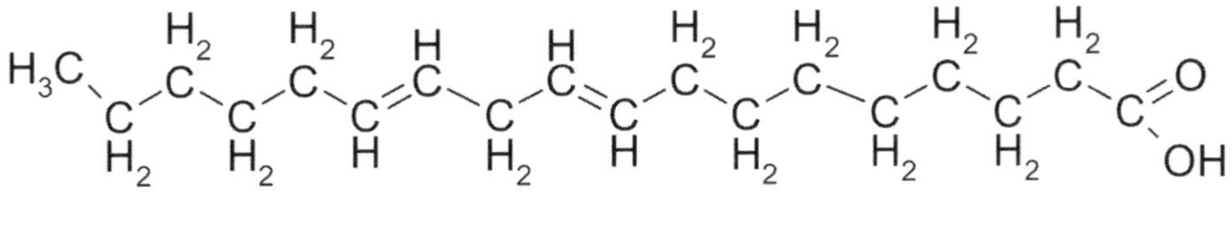

linoleic acid

Carboxylic acids containing a –COOH group at each end of the carbon chain are known as *dicarboxylic acids* or simply *diacids*. They are named by adding –dioic acid to the end of the alkyl root name. Examples include propanedioc acid, commonly known as malonic acid, and butanedioic acid, commonly known as succinic acid, both of which are being proposed as precursors in the synthesis of biobased polymers.

$$HO-\overset{\overset{\displaystyle O}{\|}}{C}-CH_2-\overset{\overset{\displaystyle O}{\|}}{C}-OH \qquad\qquad HO-\overset{\overset{\displaystyle O}{\|}}{C}-CH_2-CH_2-\overset{\overset{\displaystyle O}{\|}}{C}-OH$$

propanedioc acid   butanedioic acid

The boiling points of carboxylic acids are high because the formation of hydrogen bonded dimers causes the molecules to interact very strongly. Carboxylic acids serve both as hydrogen bond donors and hydrogen bond acceptors; thus, carboxylic acids are soluble in water. Liquid carboxylic acids have sharp, unpleasant odors.

The carboxyl group readily serves as a proton donor, thus carboxylic acids are relatively strong acids compared to alcohols and are able to react with some metals to liberate hydrogen and react with metallic carbonates to liberate carbon dioxide, but they are rather weak acids compared to mineral acids.

The reaction of carboxylic acid with a base yields the conjugate base $RCOO^-$ known as the *carboxylate* ion, which is named by replacing the -oic acid ending with -oate. The salt of a carboxylic acid is named by preceding the name of the carboxylate ion by the name of the metal ion.

$$CH_3-CH_2-CH_2-\overset{\overset{\displaystyle O}{\|}}{C}-OH + NaOH \longrightarrow CH_3-CH_2-CH_2-\overset{\overset{\displaystyle O}{\|}}{C}-O^-\ Na^+ + H_2O$$

butanoic acid   sodium butanoate

Because they are ionic, carboxylates are more soluble than their corresponding carboxylic acids. This feature is often used to separate carboxylic acids from other non-polar organic compounds.

2.3.9 Esters

The RCO unit in a carboxylic acid is called an *acyl* group. A number of carboxylic acid derivatives result by replacing the hydroxyl group that is attached to the acyl group of a carboxylic acid with other atoms or functional groups, which are known as substituents.

If an alkoxy group (R-O) or phenoxy group (Ar-O) is bonded to the acyl group, the derivative is called an ester.

$$R-\overset{\overset{\displaystyle O}{\|}}{C}-O-R' \qquad\qquad R-\overset{\overset{\displaystyle O}{\|}}{C}-O-Ar$$

ester based on   ester based on
alkoxy group   phenoxy group

Esters are named by adding the suffix -oate to the parent name of the corresponding carboxylic acid and adding a prefix appropriate to the alkoxy or phenoxy group. For example, the acyl group from butanoic acid combined with the alkoxy group derived from the methyl group yields methyl butanoate, a chemical that gives apples their aroma.

$$CH_3{-}CH_2{-}CH_2{-}\overset{\overset{\displaystyle O}{\|}}{C}{-}O{-}CH_3$$

methyl butanoate

Esters are polar molecules. However, because there is no intermolecular hydrogen bonding between the molecules, the boiling points of esters are lower than those of alcohols and carboxylic acids of comparable molecular weight. The oxygen atoms in esters can form hydrogen bonds to the hydrogen in water; thus, esters are slightly soluble in water. They are less soluble than carboxylic acids, though, because they have no hydrogen to form a hydrogen bond to the oxygen in water. Esters have pleasant fruity smells that are responsible for the aroma of many fruits.

Esters are synthesized from carboxylic acids by condensation reactions with alcohols in the presence of an acid catalyst. In these condensation reactions, known as esterification, the carboxylic acid supplies the hydroxyl group while the alcohol supplies the hydrogen to form water. According to Le Chaterlier's principle, distilling water out of the reaction mixture can increase the yield of ester. Likewise, reacting in the presence of excess alcohol can increase the ester yield.

$$R{-}\overset{\overset{\displaystyle O}{\|}}{C}{-}OH \;+\; H{-}O{-}R' \;\rightleftharpoons\; R{-}\overset{\overset{\displaystyle O}{\|}}{C}{-}O{-}R' \;+\; H_2O$$

carboxylic acid   alcohol   ester

Hydrolysis of esters, catalyzed by a strong acid, is the reverse reaction of esterification, yielding a carboxylic acid and an alcohol as products. However, if hydrolysis occurs in the presence of a strong base, the conjugate base of the carboxylic acid, known as a carboxylate ion, forms instead of the acid. This process is called saponification, from the Latin word for soap since soaps are the salts of long-chain carboxylic acids (metal carboxylates).

Triglycerides, also known as fats and oils, are a special class of esters that consist of three long-chain fatty (carboxylic) acids attached to a backbone of glycerol

(1,2,3-propanetriol). The acid fractions of triglycerides can vary in chain length and degree of saturation. Fats, which are solid or semi-solid at room temperature, have a high percentage of saturated acids, whereas oils, which are liquid at room temperature, have a high percentage of unsaturated acids. Plant-derived triglycerides are typically oils containing unsaturated fatty acids, including oleic, linoleic, and linolenic acids.

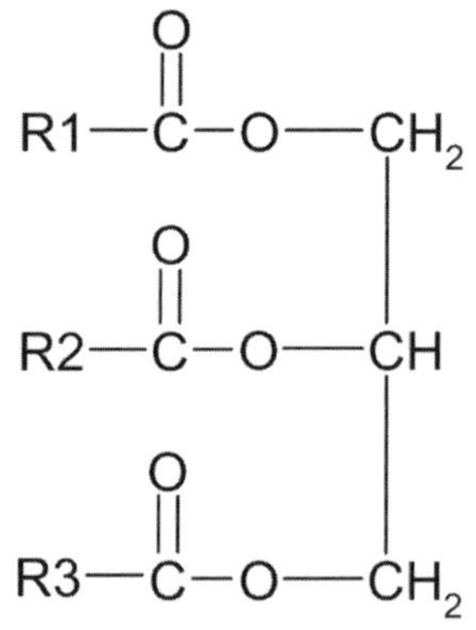

$$R1-\overset{\overset{\displaystyle O}{\|}}{C}-O-CH_2$$

$$R2-\overset{\overset{\displaystyle O}{\|}}{C}-O-CH$$

$$R3-\overset{\overset{\displaystyle O}{\|}}{C}-O-CH_2$$

triglyceride

Waxes are another class of esters consisting of fatty acids and long-chain alcohols containing an even number of carbon atoms. Waxes are low-melting solids that coat the surface of plant leaves and fruits and also the hair and feathers of some animals. Waxes are usually a complex mixture of several esters. For example, hydrolysis of beeswax yields several fatty acids and a mixture of alcohols containing 24–36 carbon atoms.

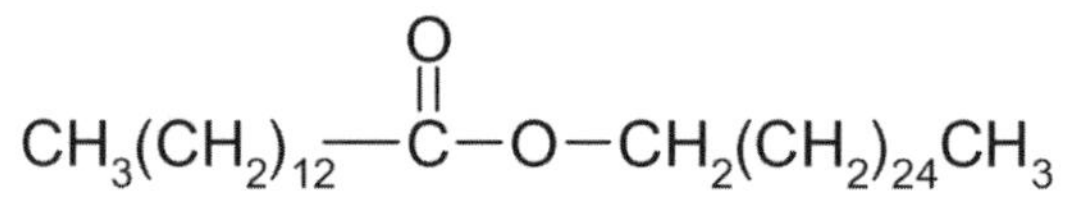

$$CH_3(CH_2)_{12}-\overset{\overset{\displaystyle O}{\|}}{C}-O-CH_2(CH_2)_{24}CH_3$$

an ester found in beeswax

Polyesters are produced by the reaction of carboxylic acids and alcohols that contain two or more functional groups. These acid-catalyzed condensation reactions yield long chains of repeating ester groups that represent an important class of synthetic polymers. One of the most important polyesters, poly(ethylene terephthalate) or PET, is produced by the reaction of 1,2-ethanediol (ethylene glycol) with benzene-1,4-dicarboxylic acid (terephthalic acid), which is spun into fibers and marketed as Dacron®.

HO-CH₂-CH₂-OH + benzene-1,4-dicarboxylic acid → poly(ethylene terephthalate)

1,2-ethanediol   benzene-1,4-dicarboxylic acid   poly(ethylene terephthalate)

2.3.10 Other Functional Groups

Other carboxylic acid derivates include *amides,* in which a substituent is linked to the acyl group by a nitrogen atom; *thioesters*, in which the substitutent is linked to the acyl group by a sulfur atom; *acyl chlorides*, in which the substituent linked to the acyl group is a chlorine atom; and *acid anhydrides*, in which two acyl groups are linked by an oxygen atom.

Organic compounds with one or more carbon-nitrogen single bonds are *amines*. Compounds with carbon-nitrogen double and triple bonds are *imines* and nitriles, respectively. Sulfur forms single bonds to carbon in two classes of compounds: *thiols* (also called mercaptans), and *thioethers* (also called sulfides). These classes structurally resemble alcohols and ethers, compounds that contain oxygen, another element in the same group of the periodic table as sulfur.

2.4 Chemistry of Plant Materials

2.4.1 Carbohydrates

Carbohydrates are polyhydroxy aldehydes, polyhydroxy ketones, or compounds that can be hydrolyzed from them. Carbohydrates, which are designated with the suffix -ose, range in size from molecules containing three carbon atoms to gigantic molecules containing thousands of carbon atoms. The smallest carbohydrates, those that cannot be hydrolyzed to smaller carbohydrate units, are called *monosaccharides.*

Carbohydrates consisting of a few monosaccharides are called *oligosaccharides.* Those consisting of two monosaccharides, which may be the same or different compounds, are called *disaccharides.* Examples include lactose and maltose. *Polysaccharides* contain thousands of covalently linked monosaccharides. *Starch* and *cellulose* are examples of *homopolysaccharides,* which are polysaccharides containing only one kind of monosaccharide. Polysaccharides that contain different kinds of monosaccharides are called *heteropolysaccharides*. Acetal or ketal bonds link the monosaccharides in oligo- and polysaccharides; these can be hydrolyzed to yield the component monosaccharides.

   Monosaccharides are classified according to their most highly oxidized functional group, either an aldehyde or a ketone group. *Aldoses* contain a single aldehyde group while *ketoses* contain a single ketone group. Monosaccharides are designated with the prefix aldo- or keto-, as appropriate, and the prefix tri-, tetra-, pent-, and hex- to indicate the number of carbon atoms. For example D-glucose and D-fructose are 6-carbon monosaccharide isomers with the molecular formula of $C_6H_{12}O_6$. However, D-glucose contains an aldehyde group as the first carbon atom in the chain, making it an aldohexose, while D-fructose contains a ketone group as the second carbon atom in the chain, making it a ketohexose.

```
        CHO                      CH₂OH
         |                         |
   H—C—OH                      C=O
         |                         |
  HO—C—H                    HO—C—H
         |                         |
   H—C—OH                     H—C—OH
         |                         |
   H—C—OH                     H—C—OH
         |                         |
      CH₂OH                     CH₂OH

    D-glucose                 D-fructose

  (an aldohexose)          (a ketohexose)
```

   D-glucose is the carbohydrate around which metabolism is based, and D-fructose is the sugar commonly found in fruits and honey. Because of the differences in functional groups, D-glucose and D-fructose qualify as functional isomers of one another.

   The monosaccharides also include *stereoisomers*—those that have the same sequence of bonded atoms but differ only in the spatial location of atoms or functional groups around the carbon chain. Of particular interest are *epimers*, isomers that differ in spatial arrangement only about a single carbon atom in the chain. D-galactose, like D-glucose, is an aldohexose with molecular formula $C_6H_{12}O_6$. They differ only in the position of the hydroxyl group attached to the fourth carbon atom (where carbon atoms are numbered starting at the end of the chain closest to the carbonyl carbon atom); thus, they are known as C-4 epimers. Although physical and chemical properties of epimers are essentially the same, biological activity can be quite different.

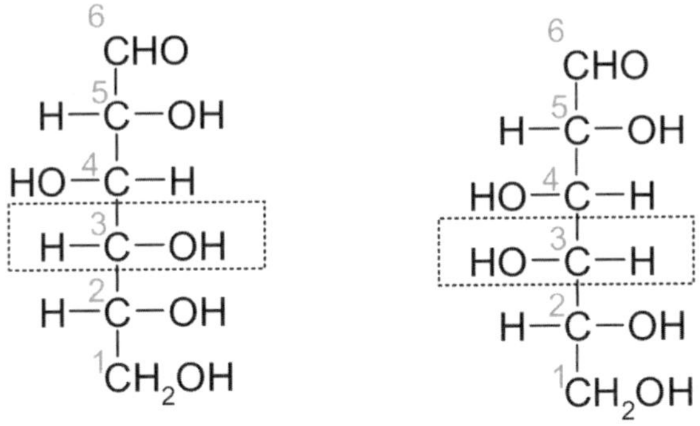

$$
\begin{array}{c}
\overset{6}{C}HO \\
H-\overset{5}{C}-OH \\
HO-\overset{4}{C}-H \\
\boxed{H-\overset{3}{C}-OH} \\
H-\overset{2}{C}-OH \\
\overset{1}{C}H_2OH
\end{array}
\qquad
\begin{array}{c}
\overset{6}{C}HO \\
H-\overset{5}{C}-OH \\
HO-\overset{4}{C}-H \\
\boxed{HO-\overset{3}{C}-H} \\
H-\overset{2}{C}-OH \\
\overset{1}{C}H_2OH
\end{array}
$$

D-glucose   D-galactose

A special instance of positional isomers among monosaccharides is based on chirality. A molecule is *chiral* if it is not superimposable on its mirror image; otherwise it is achiral. Chiral molecules have the same chemical formulas but can be arranged in two different ways corresponding to the mirror images of one another; these are known as *enantiomers*. In carbohydrate chemistry, enantiomers are designated by either the prefix D- or L-, the former being used for enantiomers that correspond to the chirality of the naturally occurring enantiomer of glyceraldehyde, a 3-carbon monosaccharide. Since monosaccharides in nature are derived from the building block D-glyceraldehyde, nearly all naturally derived monosaccharides are of the D series. Thus, the D- or L- prefix is commonly discarded when discussing monosaccharides, with the D series assumed.

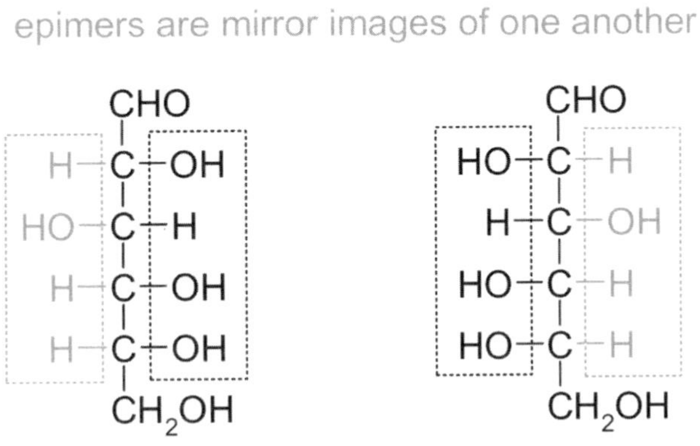

epimers are mirror images of one another

$$
\begin{array}{c}
CHO \\
H-C-OH \\
HO-C-H \\
H-C-OH \\
H-C-OH \\
CH_2OH
\end{array}
\qquad
\begin{array}{c}
CHO \\
HO-C-H \\
H-C-OH \\
HO-C-H \\
HO-C-H \\
CH_2OH
\end{array}
$$

D-glucose   L-glucose

Recall that the carbonyl group of aldehydes and ketones can react reversibly with the hydroxyl group of alcohols to form hemiacetals and hemiketals, respectively. Furthermore, if the carbonyl group and the hydroxyl groups occur in the same compound, they can react to form cyclic structures known as cyclic hemiacetals and cyclic hemiketals. Monosaccharides contain both carbonyl groups and hydroxyl groups and thus readily form cyclic structures. Indeed, hemiacetal forms of aldohexoses and aldopentoses predominate over open chain five-carbon and six-carbon aldoses while hemiketal forms of ketohexoses and ketopentoses predominate over open chain five-carbon and six-carbon ketoses. Cyclic hemiacetals and cyclic hemiketals of carbohydrates that form 5-membered rings are called furanoses; those that form 6-membered rings are called *pyranoses*. The carbon atoms are numbered starting with the carbon atom nearest to the carbon atom at the anomeric center.

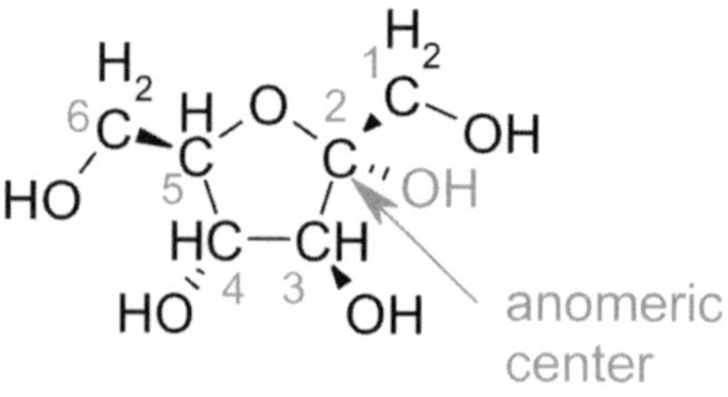

perspective structural
formula of a furanose

(α-D-fructofuranose)

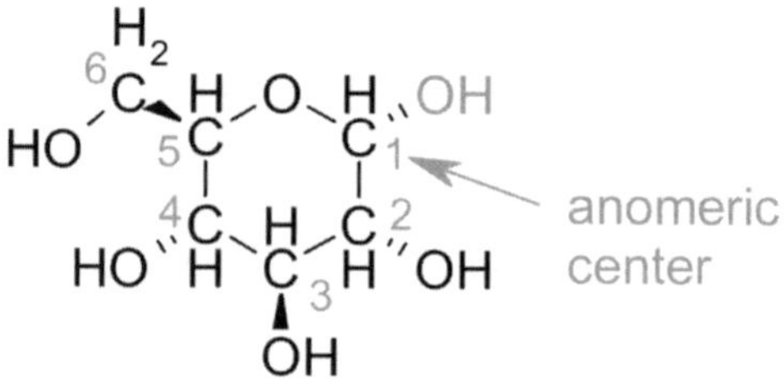

perspective structural
formula of a pyranose

(α-D-glucopyranose)

The carbon atom around which the hemiacetal or hemiketal structure forms is called the *anomeric carbon atom* or *anomeric center*. Notice that it is the only carbon atom in the monomer attached to two oxygen atoms. Also notice that the hydroxyl groups attached to the anomeric centers of the furanose and pyranose molecules illustrated above are directed below the plane of the cyclic structure (as indicated by the dashed wedge used to represent the C-OH bond). This isomeric form is designated by the prefix α-. Thus, the hemiketal form of D-fructose with a downward projecting hydroxyl group attached to the anomeric center is called α-D-fructofuranose, while the hemiacetal form of D-glucose with a downward projecting hydroxyl group attached to the anomeric center is called α-D-glucopyranose. The epimers of these monosaccharides have the hydroxyl group associated with the anomeric center projecting upwards and are designated with a prefix β-.

β-D-fructofuranose   β-D-glucopyranose

Recall that hemiacetals and hemiketals can react with alcohols to form acetals and ketals. The acetals and ketals produced from the cyclic hemiacetal and hemiketal forms of monosaccharides are called *glycosides* and the C–O bond thus formed is called the *glycosidic bond*. The group bonded to the anomeric carbon atom of a glycoside is an *aglycone*.

β-D-glucopyranose   methanol   methyl β-D-glucopyranoside

These glycosidic bonds are the basis for constructing oligosaccharides and polysaccharides from monosaccharides since they allow linkage between the anomeric center of one monosaccharide with a hydroxyl oxygen atom of a second monosaccharide. Consider the bonding of two D-glucose monosaccharides in their cyclic forms of β-D-glucopyranose. A glycosidic bond is formed between the anomeric center of the first monosaccharide (designated the C-1 carbon atom) and one of the hydroxyl carbon atoms of the second monosaccharide. If the fourth carbon atom of the second monosaccharide is involved (called the C-4' carbon atom), the resulting disaccharide is called cellobiose, the structural unit of cellulose. The IUPAC name is 4-O-(β-D-glucopyranosyl)-β-D-glucopyranose. The -syl suffix on the first monosaccharide parent name indicates that this unit is linked to the second unit by a glycosidic bond. The prefix 4-O refers to the position of the oxygen atom of the aglycone unit.

aglycone

glycosidic bond

β-D-glucopyranose   +   β-D-glucopyranose   ⇌   4–O–(β-D-glucopyranosyl)-β-D-glucopyranose
(cellobiose)   + H₂O

Notice that the reverse of this disaccharide-building reaction is hydrolysis of an oligosaccharide to release monosaccharides.

Other common disaccharides are maltose, a combination of α-D-glucopyranose and β-D-glucopyranose connected by a α-1,4'glycosidic bond; lactose, a combination of β-D-galactopyranose and β-D-glucopyranose joined by a β-1,4'glycosidic bond; and sucrose, a combination of α-D-glucopyranose and β-D-fructofuranose linked by their anomeric centers by both an α glycosidic bond on the glucose and a β glycosidic bond on the fructose.

Among the most important polysaccharides in nature are starch, cellulose, and hemicellulose. *Starch* is a polymer of α-1,4-linked glucose molecules consisting of the disaccharide maltose as the basic structural unit. Starch, an important energy source in nature, accumulates as granules in the cells of many kinds of plants. Starch occurs as both linear molecules known as *amylose,* and as branched molecules known as *amylopectin.* Different plants accumulate various proportions of amylose and amylopectin. Cellulose and hemicellulose are discussed in the next section.

2.4.2 Lignocellulose

Lignocellulose is the term used to describe the three-dimensional polymeric composites formed by plants as structural material. It consists of variable amounts of cellulose, hemicellulose, and lignin. Hardwoods (from deciduous trees), softwoods (from coniferous trees), and herbaceous material (from grasses and agricultural crops) have distinct compositions from one another, as detailed in Chapter 3.

Cellulose, a homopolysaccharide of glucose, is an important constituent of most plants. The basic building block of this linear polymer is cellobiose, a compound of two glucose molecules. The number of glucose units in a cellulose chain is known as the *degree of polymerization (DP).* The average DP for native cellulose is on the order of 10,000 although chemical pulping reduces this to the range of 500 –2,000. Cellulose molecules are randomly oriented with a tendency to form intra- and intermolecular hydrogen bonds. The strong tendency for intra- and intermolecular hydrogen bonding in cellulose results in molecular aggregation to form *microfibrils*. High packing densities result in highly ordered microfibrils known as crystalline cellulose. Low packing densities result in less ordered microfibrils known as amorphous cellulose. Crystalline cellulose is relatively inert to chemical treatment and insoluble in most solvents.

Hemicellulose is a large number of heteropolysaccharides built from hexoses (D-glucose, D-mannose, and D-galactose), pentoses (D-xylose, L-arabinose, and D-arabinose), and deoxyhexoses (L-rhamnose or 6-deoxy-L-mannose and rare L-fucose or 6-deoxy-L-galactose). Small amounts of uronic acids (4-O-methy-D-glucuronic acid, D-galacturonic acid, and D-glucuronic acid) are also present. In softwoods, the primary hemicellulose components are galactoglucomannans (glucomannan) and arabinoglucuronoxylan (xylan). Softwood hemicelluloses have more mannose and galactose units and less xylose units and acetylated hydroxyl groups than do hardwood hemicelluloses. The monosaccharides released upon hydrolysis of hemicellulose include a large fraction of pentoses as opposed to hexoses from cellulose. The chemical and thermal stability of hemicelluloses is lower than that of cellulose, presumably due to their lack of crystallinity and lower degree of polymerization, which is only 100–200.

Lignin, a phenylpropane-based polymer, is the largest non-carbohydrate fraction of lignocellulose. It is constructed of three monomers: coniferyl alcohol, sinapyl alcohol, and coumaryl alcohol, each of which has an aromatic ring with different substituents. The functional groups associated with lignin, including phenolic and alcoholic hydroxyl groups, and aldehyde (CHO-) and methoxy (CH_3O-) groups, result in highly reactive molecules. The number of various functional groups per 100 phenylpropane (C_6H_3) units is given in Table 2.5 for hardwoods and softwoods. Unlike cellulose, lignin cannot be depolymerized to its original monomers. Lignin and hemicellulose form a sheath that surrounds the cellulosic portion of the biomass. Lignin protects lignocellulose against insect attack.

Natural lignins are roughly classified according to plant source: softwood, hardwood, and grasses. Attempts to chemically liberate lignin from lignocellulose almost always produce a modified product distinct from the natural form with different physical and chemical properties. Thus, it is common to distinguish lignin by the process that liberated it: kraft or sulfate lignin from kraft pulping; alkali or

Table 2.5. Number of functional groups in natural lignin (per 100 C_6H_3 units)

Functional Group	Softwood Lignin	Hardwood Lignin
Phenolic hydroxyl (Ar-OH)	20–30	10–20
Aliphatic hydroxyl (R-OH)	115–120	110–115
Methoxyl (CH_3O-)	90–95	140–160
Aldehyde (CHO-)	20	15

soda lignin from soda processing; lignosulfonates from sulfite pulping; organosolv lignin from treating wood with alcohol solvents; and acid hydrolysis and enzymatic hydrolysis lignins from these respective processes to "*saccharify*" lignocellulose. The lignin is condensed to different degrees by these processes. Very fine mechanical milling of wood can liberate "milled wood" lignin, which is thought to be very close to natural lignin in composition and chemistry.

Plant materials also contain thousands of other chemical compounds known as *extractives*. These include resins, fats and fatty acids, phenolics, phytosterols, and other compounds, the content of which is extremely dependent on the plant species. These compounds are often classified as either *hydrophilic* or *lipophilic* depending on whether they are soluble in water or organic solvents, respectively. *Resin* is often used to describe the lipophilic extractives with the exception of phenolic substances. Extractives impart color, odor, and taste to wood. Although some extractives (lipids) are an energy source for the plant, the functions of most appear to be to protect the plant against microbiological damage or insect attacks. They are a valuable by-product of many manufacturing processes, especially in the pulp and paper industry. For example, southern pines, favored in pulp making, have particularly high content of extractives, which are recovered as crude turpentine and raw tall oil.

The cell walls of both woody and herbaceous biomass consist of lignocellulose. These cells can be classified into two broad categories: *prosenchyma* cells, which are long, thin cells, with flattened or tapered closed ends; and *parenchyma* cells, which are short, rectangular cells. Prosenchyma cells are on the order of 1–6 mm in length with widths of 20–50 μm. Parenchyma cells are shorter than 0.2 mm with widths of 2–50 μm.

Softwoods consist primarily of prosenchyma cells, about 90% of the total, referred to as *tracheids,* which both transport fluids through the plant and support the plant. These are long, strong fibers ideal for pulp and paper applications. Parenchyma cells provide storage of nutrients in softwoods.

Hardwoods consist of a much smaller fraction of prosenchyma cells, about 55% of the total, along with a significant fraction of parenchyma cells, about 20%, and cells intermediate in size to prosenchyma and parenchyma cells known as *vessel* cells. In hardwoods, prosenchyma cells provide structural support, parenchyma cells provide both transport of fluids and nutrient storage, and vessel cells serve primarily for transport. Herbaceous plant material more closely resembles hardwoods than softwoods.

Further Reading

Engineering Thermodynamics

Moran, M. and H. Shapiro. 1999. *Fundamentals of Engineering Thermodynamics*, Fourth Edition. New York: Wiley.

Organic Chemistry

Ouellette, R. J. 1998. *Organic Chemistry: A Brief Introduction*, Second Edition. Englewood Cliffs: Prentice Hall.

Plant Chemistry

Alén, R. 2000. "Structure and Chemical Composition of Wood." *Forest Products Chemistry*. Per Stenius, ed. Helsinki: Fapet Oy.

Ouellette, R.J. 1998. *Organic Chemistry: A Brief Introduction*, Second Edition. Englewood Cliffs: Prentice Hall.

Rowell, R.M., R. A. Young, and J. R. Rowell. 1997. *Paper and Composites from Agro-based Resources*. CRC Press.

Wayman, M. and S. Parekh. 1990. *Biotechnology of Biomass Conversion: Fuels and Chemicals from Renewable Resources*. Philadelphia: Open University Press.

Problems

2.1 Fermentation of starch and sugar crops yields about 10% molar volume of ethanol in water. Distillation of this mixture to remove the water and produce pure ethanol is an expensive and energy-intensive process. To reduce fuel costs, some researchers have proposed partial distillation to a 50% mixture of ethanol and water, a mixture that could be used to power a high temperature fuel cell. Calculate the enthalpy change in oxidizing this mixture in air if the reactants are at 298 K and the products are at 1200 K.

2.2 A 5 MW gas turbine power plant is reported to have a thermodynamic efficiency of 35%. Assume products of reaction exit at 750 K with water in the gaseous phase.

   (a) What is the corresponding heat rate of this power cycle?

   (b) How much methane (kg/h) is required to power this plant?

2.3 On a dry basis, wood chips have a nominal enthalpy of reaction of 20 MJ/kg. Adiabatic flame temperature is defined as the temperature achieved by combustion products if Q_{CV} and W_{CV} in Eq. 2.11 are set to zero and reactants are assumed to be at 25°C.

   (a) Calculate the adiabatic flame temperature for dry wood chips.

   (b) Calculate the adiabatic flame temperature for wood chips containing 30 wt-% moisture.

2.4 Terpenes are an important class of "natural" compounds found in many kinds of plants. What is their relationship to conjugated dienes? What bonding arrangement characterizes conjugated dienes?

2.5 The petrochemical industry can produce ethanol from fossil resources while the corn processing industry can produce ethylene from a fermentation product. What reversible reaction is involved?

2.6 Pyrolytic decomposition of biomass yields phenols as one product. Fermentation of monosaccharides produces alcohols. How are phenols similar to alcohols? How are they different?

2.7 Plant oils are triglycerides. Write the reaction of one triglyceride molecule with three methanol molecules to release the triol backbone of the triglyceride to form three molecules of methyl ester. This reaction, known as transesterification, is the process for manufacturing biodiesel.

2.8 Describe how monosaccharides are classified.

2.9 While starch is readily hydrolyzed to glucose, cellulose is more difficult to hydrolyze. What are the similarities between starch and cellulose? What accounts for the differences in the susceptibility of these carbohydrates to hydrolysis?

2.10 What are the differences in cellular structure between hardwoods and softwoods?

CHAPTER

3

The Biorenewable Resource Base

3.1 Defining the Resource

Biorenewable resources, sometimes referred to as biomass, are organic materials of recent biological origin. This definition is deliberately broad with the intent of distinguishing fossil fuel resources from the wide variety of organic materials that arise from the biotic environment. Biorenewable resources are generally classified as either *wastes* or *dedicated energy crops*.

A waste is a material that has been traditionally discarded because it has no apparent value or represents a nuisance or even a pollutant to the local environment. Clearly, if so-called wastes from one process were utilized as feedstock in another process, a more appropriate name would be co-products. For example, oat-processing plants often generate enormous quantities of agricultural residues in the form of hulls that are currently viewed as wastes. If economically converted into process heat, electricity, liquid fuels, or chemicals, they would be considered a co-product rather than a waste stream. This holistic approach to manufacturing, in which all the outputs from one process become the inputs to other processes, is known as *industrial ecology*. However, the word "wastes" remains a convenient moniker for "low value co-products" and will be used in this book.

Dedicated energy crops are plants grown specifically for production of biobased products; that is, for purposes other than food or feed. The term was originally coined to describe woody or herbaceous plants grown for their high yields of lignocellulosic material, which can be burned in a power plant to produce electricity or hydrolyzed to release fermentable sugars suitable for the production of transportation fuels. However, not all dedicated energy crops are grown for fuels and energy (they might be used for production of commodity chemicals or natural fibers), and not all fuels and energy products are derived from lignocellulosic crops (indeed, fuel ethanol is currently produced from corn starch in the United States and sugar cane in Brazil). Thus, the term "dedicated energy crop" is something of a misnomer, but it has wide usage and is understood to mean crops grown specifically as a source of carbon and energy for the manufacture of biobased products.

3.1.1 Waste Materials

Categories of *waste materials* that qualify as biorenewable resources include agricultural residues, yard waste, municipal solid waste, food processing waste, and even manure. Agricultural residues are simply that part of a crop discarded after harvest such as corn stover (husks and stalks), rice hulls, wheat straw, bagasse (fibrous material remaining after the milling of sugar cane), grapevine prunings, and almond shells, to name a few. Yard waste is an urban biomass crop: grass clippings, leaves, and tree trimmings.

Municipal solid waste (MSW) is whatever is thrown out in the garbage, not all of which is suitable as biomass feedstock. In some communities, yard waste may constitute up to 18% of MSW, although a growing number of communities have ordinances against disposal of yard waste with garbage in an effort to conserve landfill space. In communities where yard waste is excluded, the important components of MSW are paper (50%), plastics and other fossil fuel-derived materials (20%), and food wastes (10%). Non-flammable materials (glass and metal) represent 20% of MSW.

Food processing waste is the effluent from a wide variety of industries ranging from breakfast cereal manufacturers to alcohol breweries. These wastes may be dry solids or watery liquids. Sewage represents a source of chemical energy and is often converted into electric power at municipal wastewater treatment plants. The recent concentration of animals into giant livestock facilities has led to calls to treat animal wastes in a manner similar to that for human wastes. Consequently, many strategies for manure management integrate waste treatment with heat and power generation.

Waste materials share few common traits other than the difficulty of characterizing them because of their variable and complex composition. Municipal solid waste is the leavings of thousands of households and industries that yield a feedstock that may be easy to process one day and difficult the next. Yard wastes show seasonal variations in quantity and composition: the spring brings high moisture grass clippings that are replaced by dry leaves in the autumn. Waste streams from food processing plants, on the other hand, may be relatively invariant in composition but contain a wide assortment of complex organic compounds that are not amenable to a single conversion process. Thus, waste biomass presents special problems to engineers who are charged with converting this sometimes-unpredictable feedstock into reliable power or high quality fuels and chemicals.

The major virtue of waste materials is their low cost. By definition waste materials have little apparent economic value and often can be acquired for little more than the cost of transporting the material from their point of origin to a processing plant. Increasing costs for solid waste disposal and sewer discharges, and restrictions on landfilling certain kinds of wastes allow some wastes to be acquired at negative cost; that is, a biorenewable resource processing plant is paid by a company seeking to dispose of a waste stream. For this reason many of the most economically attractive opportunities in biorenewable resources involve waste feedstocks. For example, the seed corn industry, which sells seed grown specifically for planting new crops, has an annual waste disposal problem. Seed for which germination cannot be guaranteed after a certain period of storage is taken off the market. This seed cannot be sold for animal feed or even landfilled because the seed is treated with fungicide. Seed corn companies often pay brokers to accept this obsolete seed, and they in turn sell it as an inexpensive fuel for boilers and cement kilns.

As demand for these new-found feedstocks increases those that generate them come to view themselves as suppliers and may demand payment for their waste stream: a negative feedstock cost becomes a positive one. Such a situation developed in the California biomass power industry during the 1980s. Concerns about air pollution in California led to restrictions on open-field burning of agricultural residues, a practice designed to control infestations of pests. With no means for getting rid of these residues an enormous reserve of biomass feedstocks materialized. These feedstocks were so inexpensive independent power producers recognized that even small, inefficient power plants using these materials as fuel would be profitable. A number of plants were constructed and operated on agricultural residues. Eventually, the feedstock producers had plant operators bidding up the cost of their once valueless waste material. In the end, many of these plants were closed because of the escalating cost of fuel.

3.1.2 Dedicated Energy Crops

Dedicated energy crops are defined as plants grown specifically for applications other than food or feed. It is important to note that firewood obtained from cutting down an old-growth forest does not constitute a dedicated energy crop. A dedicated energy crop is planted and harvested periodically. Harvesting may occur on an annual basis, as with sugar beets or switchgrass, or on a 3- to 10-year cycle, as with certain strains of fast-growing trees such as hybrid poplar or willow. The cycle of planting, harvesting, and regrowing over a relatively short time period assures that the resource is used in a sustainable fashion; that is, the resource will be available for future generations.

Dedicated energy crops can fulfill one or more market niches. In some instances, the whole plant is used as feedstock for production of electricity and/or liquid fuels. Such is the case when trees are grown and harvested specifically as boiler fuel for steam power plants. Another possibility is that a variety of co-products are coaxed from a single crop. For example, alfalfa has been evaluated for its potential to yield both energy and feed from a single crop. The high protein leaves would be removed after harvesting and processed into animal feed while the fibrous stems would be used as fuel in a gasification power plant. The least desirable and most wasteful scenario for dedicated energy crops is extraction of the highest-value portion of the crop for conversion into biobased product and discarding the rest of the plant as waste. Milling sugar cane to extract sugar for fermentation to ethanol and discarding the rest of the plant material (known as bagasse) is an example of a conversion process that is wasteful of a biorenewable resource. A better strategy for utilizing this resource would be to extract the sugar as food and to convert the bagasse to commodity chemicals or electric power.

Dedicated energy crops contain significant quantities of one or more of four important energy-rich components: oils, sugars, starches, and lignocellulose (fiber). Crops rich in the first three have historically been grown for food and feed: oils from soybeans, nuts, and grains; sugars from sugar beets, sorghum, and sugar cane; and starches from corn and cereal crops. Oil, sugars, and starches are easily metabolized. On the other hand, lignocellulose is indigestible by humans although certain domesticated animals with specialized digestive tracts are able to break down the polymeric structure of lignocellulose and use it as an energy source. From this discussion it might appear that the best strategy for developing biomass resources is to grow crops rich in oils, sugars, and starches. However, even for "oil crops" or "starch crops" the largest single constituent is invariably lignocellulose, which is the structural (fibrous) material of the plant: stems, leaves, and roots. If oils, sugars, and starches are harvested and the lignocellulose is left behind as an agricultural residue rather than used as fuel or feedstock, the greatest portion of the biomass crop remains in the field.

Not only should lignocellulose be valued, there is good reason to maximize its production at the expense of lipids and simple carbohydrates if energy production or commodity chemicals are the primary purpose for growing the crop. Research has shown that energy yields (megajoules per hectare per year) are usually greatest for plants that are mostly "roots and stems;" in other words, plant resources are directed toward the manufacture of lignocellulose rather than oils, sugars, and starches. As a result, there has been a bias toward development of dedicated energy crops that focus on lignocellulosic biomass, a bias that is reflected in the discussion that follows. Lignocellulosic crops are conveniently divided into *herbaceous energy crops (HEC)* and *short-rotation woody crops (SRWC)*.

3.1.2.1 Herbaceous Energy Crops

Herbaceous crops are plants that have little or no woody tissue. The above ground growth of these plants usually lives for only a single growing season. However, herbaceous crops include both annuals and perennials. Annuals die at the end of a growing season and must be replanted in the spring. Perennials die back each year in temperate climates but reestablish themselves each spring from rootstock. Both annual and perennial herbaceous energy crops are harvested on at least an annual basis, if not more frequently, with yields averaging 5.5–11 Mg/ha/yr, with maximum yields between 20 and 25 Mg/ha/yr in temperate regions. As with trees, yields can be much higher in tropical and subtropical regions.

Among the many species of herbaceous plants that are potentially suitable as dedicated energy crops, recent development work has focused on grasses because of their high yields of lignocellulose. Grasses are conveniently classified as either thick-stemmed or thin-stemmed. Thick-stemmed grasses, which include annual and perennial varieties, are indigenous to the tropics. The most familiar examples are sugarcane and energy cane (*Sacharum* spp.) and napiergrass (*Pennisetum purpureum*) among the perennials and corn (*Zea mays*) and forage sorghum (sorghum, sudangrass, and sorghum × sudangrass, now all classified as sorghum bicolor) among the annuals.

Harvesting of thick-stemmed perennials such as sugar cane is a labor-intensive activity, even with mechanized harvesting equipment. Cost-effective harvesting of thick-stemmed perennials as HEC would probably be by forage harvesters followed by storage as silage. The same is true of many of the thick-stemmed annuals although dry corn stalks can be baled readily.

Thin-stemmed grasses include many perennial and annual species. These are conveniently classified as either cool-season grasses, which grow more vigorously in the spring and fall, and warm-season grasses, which grow most actively during the summer. Familiar perennial cool-season grasses include reed canarygrass (*Phalaris arundinacea*), timothy (*Phleum pratense*), and tall fescue (*Festuca arundinacea*). Examples of warm-season grasses are switchgrass (*Panicum virgatum*), big bluestem (*Andropogon gerardii*), and eastern gamagrass (*Tripsacum dactyloides*).

The thin-stemmed perennials are particularly attractive as HEC because they can be harvested with conventional hay equipment. They are less susceptible than the thick-stemmed grasses to lodging (falling over on one another as the plants become tall). This is important because it allows the plants to be harvested at the end of the growing season when valuable nutrients have translocated to roots. Perennials, as a rule, are more drought resistant than annuals, require less weed control, and are less likely to erode soils. Warm-season, thin-stemmed grasses are the leading candidates for HEC. They are more drought resistant than cool-season grasses and are efficient users of nutrients.

Herbaceous crops more closely resemble hardwoods in their chemical properties than they do softwoods. Their low lignin content makes them relatively easy to delignify, which improves accessibility of the carbohydrate in the lignocellulose. The hemicellulose in herbaceous crops contains mostly xylan, which is highly susceptible to acid hydrolysis. As a result, agricultural residues are susceptible to microbial degradation, destroying their processing potential in a matter of days if exposed to the elements. Herbaceous crops have relatively high silica content compared to woody crops, which can present problems during processing.

3.1.2.2 Short-Rotation Woody Crops

Short-rotation woody crop (SRWC) is the term used to describe woody biomass that is fast growing and suitable for use in dedicated feedstock supply systems. Desirable SRWC candidates display rapid juvenile growth, wide site adaptability, and pest and disease resistance. Woody crops grown on a sustainable basis are harvested on a rotation of 3–10 years.

Woody crops include hardwoods and softwoods. Hardwoods are trees classified as angiosperms, which are also known as flowering plants. Examples include willow, oak, and poplar. Hardwoods can resprout from stumps, a process known as coppicing, which reduces their production costs compared to softwoods. Advantages of hardwoods in processing include high density for many species; relative ease of delignification and accessibility of wood carbohydrates; the presence of hemicellulose high in xylan, which can be removed relatively easily; low content of ash, particularly silica, compared to softwoods and herbaceous crops; and high acetyl content compared to most softwoods and herbaceous crops, which is an advantage in the recovery of acetic acid. Hardwood lignin is less condensed (that is, it has a lower degree of polymerization) than softwood and contains a greater methoxyl content, which explains why it was at one time the preferred choice for the destructive distillation of wood to produce methanol. Hardwood lignin becomes plastic at lower temperatures than for softwood lignin.

Softwoods are trees classified as gymnosperms, a group that encompasses most trees known as evergreens. Examples include pine, spruce, and cedar. Softwoods are generally fast growing but their carbohydrate is not as accessible for chemical processing as the carbohydrates in hardwood. Since softwoods have considerable value as construction lumber and pulpwood, they are more readily available as waste material in the form of logging and manufacturing residues than are hardwoods. Logging residues, consisting of a high proportion of branches and tops, contain considerable high-density compression wood, which is not easily delignified. Logging residues are more suitable as boiler fuel or other thermochemical treatments than as feedstock for chemical or enzymatic processing.

Development of dedicated feedstock supply systems has focused on several hardwood species, including poplar (*Populus* spp.), willows (*Salix* spp.) silver maple (*Acer saccharinum*), sweetgum (*Liquidambar styraciflua*), sycamore (*Platanus occidentalis*), black locust (*Robinia pseudoacacia*), and *eucalyptus*. Trees of potential regional importance in the United States include alders (*Alnus* spp.), mesquite (*Prosopis* spp.), and the Chinese Tallow (*Sapium sebiferum*).

Hybrid poplar and eucalyptus are most promising for the United States because of high growth rates averaging between 10 and 17 Mg/ha/yr, depending upon geographic location, with maximum yields between 15 and 43 Mg/ha/yr. In the United States, hybrid poplar has a wider range than eucalyptus, which is limited to southern Florida, California, and Hawaii. Hybrid poplar is also attractive because it is easily propagated from either stem cuttings or tissue culture.

3.2 Properties of Biomass

Evaluation of biomass resources as potential feedstocks generally requires information about plant composition, heating value, production yields, and bulk density. Compositional information can be reported in terms of organic components, proximate analysis, or ultimate analysis.

Analysis in terms of organic components reports the kinds and amounts of plant chemicals including proteins, oils, sugars, starches, and lignocellulose (fiber). In the case of dedicated energy crops, engineers are particularly interested in the ligno-cellulosic component and how it is partitioned among cellulose, hemicellulose, and lignin. Table 3.1 gives the organic components of several types of dedicated energy crops while Table 3.2 includes the organic components of common sugar and starch crops. This information is particularly useful in designing biological processes that convert plant components into commodity chemicals and trans-portation fuels.

Proximate analysis is important in developing thermochemical conversion processes for biomass. Proximate analysis reports the yields (percent mass basis) of various products obtained upon heating the material under controlled conditions; these products include moisture, volatile matter, fixed carbon, and ash. Since moisture content of biomass is so variable and can be easily determined by gravimetric

Table 3.1. Organic components of lignocellulosic crops (dry basis)

Feedstock	Cellulose (wt-%)	Hemicellulose (wt-%)	Lignin (wt-%)	Other* (wt-%)
Bagasse	35	25	20	20
Corn stover	53	15	16	16
Corncobs	32	44	13	11
Wheat straw	38	36	16	10
Wheat chaff	38	36	16	11
Short rotation woody crops	50	23	22	5
Herbaceous energy crops	45	30	15	10
Waste paper	76	13	11	0

Sources: S. R. Bull, 1991, "The U.S. Department of Energy Biofuels Research Program," in *Energy Sources,* 13:433-442; M. Wayman and S. Parekh, 1990, *Biotechnology of Biomass Conversion: Fuels and Chemicals from Renewable Resources* (Philadelphia: Open University Press).

* Includes proteins, oils, and mineral matter such as silica and alkali.

Table 3.2. Organic components of starch & sugar crops (dry basis)

Feedstock	Protein (wt-%)	Oil (wt-%)	Starch (wt-%)	Sugar (wt-%)	Fiber (wt-%)
Corn grain	10	5	72	<1	13
Wheat grain	14	<1	80	<1	5
Jerusalem artichoke	<1	<1	<1	75	25
Sugar cane	<1	<1	<1	50	50
Sweet sorghum	<1	<1	<1	50	50

Source: M. Wayman and S. Parekh, 1990, *Biotechnology of Biomass Conversion: Fuels and Chemicals from Renewable Resources* (Philadelphia: Open University Press).

methods (weighing, heating at 100°C, and reweighing), the proximate analysis of biomass is commonly reported on a dry basis. Volatile matter is that fraction of biomass that decomposes and escapes as gases upon heating a sample at moderate temperatures (about 400°C) in an inert (non-oxidizing) environment. Knowledge of volatile matter is important in designing burners and gasifiers for biomass. The remaining fraction is a mixture of solid carbon (fixed carbon) and mineral matter (ash), which can be distinguished by further heating the sample in the presence of oxygen: the carbon is converted to carbon dioxide leaving only the ash. Table 3.3 contains the proximate analysis (dry basis) of a wide range of biomass materials.

Ultimate analysis is simply the (major) elemental composition of the biomass on a gravimetric basis: carbon, hydrogen, oxygen, nitrogen, sulfur, and chlorine along with moisture and ash. Table 3.3 contains the ultimate analysis of several biomass materials on a dry basis. Sometimes this information is presented on a dry, ash-free (daf) basis. This information is very important in performing mass balances on biomass conversion processes. Evident from Table 3.3 is the relatively high oxygen content of biomass (typically 40–45 wt-%).

Of course, from ultimate analyses, the molecular formulas can be worked out. In many instances, a generic molecular formula based on one mole of carbon is convenient for performing mass balances on a process. For example, cellulose and starch have the generic molecular formula $CH_{1.7}O_{0.83}$, hemicellulose can be represented by $CH_{1.6}O_{0.8}$, and wood is $CH_{1.4}O_{0.66}$.

Heating value is the net enthalpy released upon reacting a particular fuel with oxygen under isothermal conditions (the starting and ending temperatures are the same). If water vapor formed during reaction condenses at the end of the process, the latent enthalpy of condensation contributes to what is known as the *higher heating value (HHV)*. Otherwise, the latent enthalpy does not contribute and the *lower heating value (LHV)* prevails. These measurements are typically performed in a bomb calorimeter and yield the higher heating value for the fuel. Table 3.3 reports higher heating values for several biomass materials. Heating values of biomass can be conveniently estimated from the percent of carbon in the biomass on a dry basis using the empirical relationship:

HHV in MJ/dry kg = 0.4571 × (% C on dry basis) − 2.70   (3.1)

The inorganic constituents of biomass are important to different extents, depending on the conversion process under consideration. No comprehensive information on inorganic constituents will be provided here, although such information can be found in the literature on biomass. However, knowledge of the alkali metal content of biomass (that is, potassium and sodium) can be very important if the fuel is to be used in combustors. Experience in burning biomass reveals that excessive alkali salts in biomass, which are particularly concentrated in fast-growing biomass, can lead to ash fouling of boiler tubes, as described in Chapter

Table 3.3. Thermochemical properties of selected biomass

Biomass	HHV (dry) (MJ/kg)	Proximate Analysis (% wt., dry)			Ultimate Analysis (% wt., dry)						
		Volatile	Ash	Fixed C	C	H	O	N	S	Cl	Ash
Alfalfa straw	18.45	72.60	7.25	20.15	46.76	5.40	40.72	1.00	0.02	0.03	6.07
Almond shells	19.38	73.45	4.81	21.74	44.98	5.97	42.27	1.16	0.02		5.60
Black locust	19.71	80.94	0.80	18.26	50.73	5.71	41.93	0.57	0.01	0.08	0.97
Cedar (western red)	20.56	86.50	0.30	13.20							
Corncobs	18.77	80.10	1.36	18.54	46.58	5.87	45.46	0.47	0.01	0.21	1.40
Corn stover	17.65	75.17	5.58	19.25	43.65	5.56	43.31	0.61	0.01	0.60	6.26
Corn grain	17.20	86.57	1.27	12.16	44.00	6.11	47.24	1.24	0.14		1.27
Douglas fir	20.37	87.30	0.10	12.60	50.64	6.18	43.00	0.06	0.02		0.01
Food waste	7.59				17.93	2.55	12.85	1.13	0.06	0.38	5.10
Grape vines		80.10	2.20	17.70							
Hemlock (western)	19.89	87.00	0.30	12.70	50.40	5.80	41.40	0.10	0.10		2.20
Maize straw	17.36				47.09	5.54	39.79	0.81	0.12		5.77
Manure (cattle, fresh)	19.87	76.30	12.00	11.70	45.40	5.40	31.00	1.00	0.30		15.90
Municipal solid waste	19.47				47.60	6.00	32.90	1.20	0.30		12.00
Oak bark	19.05	83.30	2.10	14.60	49.70	5.40	39.30	0.20	0.10		5.30
Orchard prunings	20.02	82.54	0.29	17.17	49.20	6.00	43.20	0.25	0.04	0.01	1.38
Ponderosa pine	19.38	82.32	1.33	16.35	49.25	5.99	44.36	0.06	0.03	0.10	0.30
Hybrid poplar	20.72	79.72	0.36	19.92	48.45	5.85	43.69	0.47	0.01	0.02	1.43
Redwood (combined)	17.40				50.64	5.98	42.88	0.05	0.03	0.57	0.40
Refuse-derived fuel	16.14	65.47	17.86	16.67	42.50	5.84	27.57	0.77	0.48	0.12	22.17
Rice hulls	16.28	69.33	13.42	17.25	40.96	4.30	35.86	0.40	0.02	0.34	18.34
Rice straw (fresh)	15.40				41.78	4.63	36.57	0.70	0.08		15.90
Sorghum stalks	17.39	72.75	8.65	18.60	40.00	5.20	40.70	1.40	0.20	0.13	12.50
Sudan grass	17.33	73.78	11.27	14.95	44.58	5.35	39.18	1.21	0.08	0.12	9.47
Sugar cane bagasse	18.64	81.36	3.61	15.03	44.80	5.35	39.55	0.38	0.01	0.03	9.79
Switchgrass	16.82				47.45	5.75	42.37	0.74	0.08		3.50
Vineyard prunings	20.18	78.28	0.56	21.16	48.00	5.70	39.60	0.86	0.08	0.03	1.41
Walnut shells	16.02		22.40		49.98	5.71	43.35	0.21	0.01		0.71
Water hyacinth	17.51	71.30	8.90	19.80	41.10	5.29		1.96	0.41	0.03	
Wheat straw	19.95	83.17	0.25	16.58	43.20	5.00	39.40	0.61	0.11	0.28	11.40
White fir	16.30	66.04	20.37	13.59	49.00	5.98	44.75	0.05	0.01	0.01	0.20
Yard waste					41.54	4.79	31.91	0.85	0.24	0.30	20.37

Sources: Various.
Note: HHV = higher heating value.

67

Table 3.4. Alkali content of biomass

	Heating Value (MJ/kg)	Ash in Fuel (%)	Alkali in Ash (%)
Hybrid poplar	19.0	1.9	19.8
Pine chips	19.9	0.7	3.0
Tree trimmings	18.9	3.6	16.5
Urban wood waste	19.0	6.0	6.2
White oak	19.0	0.4	31.8
Almond shells	17.6	3.5	21.1
Bagasse, washed	19.1	1.7	12.3
Rice straw	15.1	18.7	13.3
Switch grass	18.0	10.1	15.1
Wheat straw	18.5	5.1	31.5

Source: T. R. Miles, Sr., T. R. Miles, Jr., L. Baxter, B. Jenkins and L. Oden, "Alkali deposits found in biomass power plants," Summary Report, National Renewable Energy Laboratory, NREL Subcontract TZ-2-11226-1, April 15, 1995.

6. Table 3.4 contains information on alkali in ash for selected biomass materials useful in designing biomass combustion systems.

Bulk density is determined by weighing a known volume of biomass that is packed or baled in the form anticipated for its transportation or use. Clearly, solid logs will have higher bulk density than the same wood chipped. Bulk density will be an important determinant of transportation costs and the size of fuel storage and handling equipment. *Volumetric energy content* is also important in transportation and storage issues. Volumetric energy content, which is simply the enthalpy content of fuel per unit volume, is calculated by multiplying the higher heating value of a fuel by its bulk density. Table 3.5 compares bulk densities and volumetric energy contents of various liquid and solid fuels. The cost of collecting large quantities of biomass can be significant. Wood or other biomass resources must generally be produced within a 50-mile radius of the processing plant to be economical, given the high transportation costs and low densities of biomass.

For many applications, including production of paper and composite materials, the length of fibers in biomass is an important property. Fiber lengths for several kinds of agricultural and forestry biomass are listed in Table 3.6. Extraction of plant fibers is discussed in Chapter 8.

3.3 Yields of Biomass

Planning a biomass conversion facility requires estimates of the total amount of land that must be put into production of biomass crops and how far crops must be transported to a facility. Thus, the annual yield of biomass crops (kilograms per hectare) is important information for an engineer working on such a project.

Table 3.5. Density and volumetric energy content of various solid and liquid fuels

Fuel	Density (kg/m³)	Volumetric Energy Content (GJ/m³)
Ethanol	790	23.5
Methanol	790	17.6
Biodiesel	900	35.6
Pyrolysis oil	1280	10.6
Gasoline	740	35.7
Diesel fuel	850	39.1
Agricultural residues	50–200	0.8–3.6
Hardwood	280–480	5.3–9.1
Softwood	200–340	4.0–6.8
Baled straw	160–300	2.6–4.9
Bagasse	160	2.8
Rice hulls	130	2.1
Nut shells	64	1.3
Coal	600–900	11–33

Sources: K. Grohmann, C. E. Wyman and M. E. Himmel, 1992, "Potential for Fuels from Biomass and Wastes," in *Emerging Technologies for Material and Chemicals from Biomass*, R. M. Rowell, T. P. Schultz and R. Narayan, eds., ACS Symp. Series 476 (Washington, D.C.: American Chemical Society); E. D. Larson, P. Svenningsson and I. Bjerle, 1989, "Biomass Gasification for Gas Turbines Power Generation," in *Electricity*, Johansson, Bodlund and William, eds (Lund, Sweden: Lund University Press).

Note: GJ = Gigajoules.

Table 3.6. Dimensions of some common lignocellulosic fibers

Type of Fiber	Fiber Dimension (mm)		
	Length	Average Length	Width
Cotton	10–60	18	0.02
Flax	5–60	25–30	0.012–0.027
Hemp	5–55	20	0.025–0.050
Manila hemp	2.5–12	6	0.025–0.040
Bamboo	1.5–4	2.5	0.025–0.040
Esparto	0.5–2	1.5	0.013
Jute	1.5–5	2	0.02
Corn stalks	1.0–1.5	—	0.02
Rice straw	0.65–3.48	—	0.005–0.014
Wheat straw	1.5	—	0.015
Deciduous wood	1–1.8	—	0.03
Coniferous wood	3.5–5	—	0.025

Sources: R. M. Rowell, 1992, "Opportunities for Lignocellulosic Materials and Composites," ACS Symposium Series 476:12–27; R. M. Rowell, R. A. Young and J. R. Rowell, eds, 1997, *Paper and Composites from Agro-Based Resources*, CRC Press.

Unfortunately, yield information does not lend itself to tabulation, since it depends on so many variables: plant variety, crop management (fertilization and pest control), soil type, landscape, climate, weather, and water drainage. Table 3.7 has been

Table 3.7. Nominal annual yields of biomass crops*

Biomass Crop	Geographical Location	Annual Yield (kg/ha)
Corn: grain	North America	7,000
Corn: cobs	North America	1,300
Corn: stover	North America	8,400
Jerusalem artichoke: tuber	North America	45,000
Jerusalem artichoke: sugar	North America	6,400
Sugar cane: crop	Hawaii	55,000
Sugar cane: sugar	Hawaii	7,200
Sugar cane: bagasse (dry)	Hawaii	7,200
Sweet sorghum: crop	Midwest U.S.	38,000
Sweet sorghum: sugar	Midwest U.S.	5,300
Sweet sorghum: fiber (dry)	Midwest U.S.	4,900
Switchgrass	North America	14,000
Hybrid poplar	North America	14,000
Wheat: grain	Canada	2,200
Wheat: straw	Canada	6,000

Source: M. Wayman and S. Parekh, 1990, *Biotechnology of Biomass Conversion: Fuels and Chemicals from Renewable Resources* (Philadelphia: Open University Press).

* Includes moisture content at harvest unless otherwise noted.

included to give an idea of the kinds of yields that might be expected in various geographical locations for dedicated energy crops that have been widely studied. Site-specific information will require discussions with state extension agents and local agronomists in combination with field trials in advance of detailed plant design.

Yields of agricultural residues are conveniently tabulated in terms of *residue factors*, defined as the ratio of dry weight of residue to the grain weight at field moisture. Thus, the weight of residue available per hectare of crop can be estimated by multiplying the residue factor by the grain yield in kilograms per hectare. Average residue factors are tabulated in Table 3.8.

Manure is a relatively small biomass resource compared to dedicated energy crops or agricultural residues. However, growing concerns about the environmental impact of manure from large concentrations of animals in many modern agricultural operations make manure a potential biomass resource. Manure production rates for various kinds of livestock and poultry are listed in Table 3.9.

3.4 Size of Resource Base

The most immediately available and economically attractive biorenewable resources are agricultural and forestry residues and municipal solid wastes. As shown in Table 3.10, agricultural and forestry residues can each supply on the

Table 3.8. Agricultural residue factors for various grain crops

Crop	Residue Factor*
Barley	1.5
Corn (<95 bu/ac)	1.0
Corn (>95 bu/ac)	1.5
Cotton	1.5
Oats	1.4
Rice	1.5
Rye	1.5
Sorghum	1.5
Soybeans	1.5
Wheat, spring	1.3
Wheat, winter	1.7

Source: Heid, W. J., Jr., 1984, "Turning Great Plains Crop Residues and Other Products into Energy," *Agricultural Economic Report No. 523.* (Washington, D.C.: U.S. Dept. of Agriculture).

*These factors are ratios of dry weight of residue to the grain weight at field moisture.

Table 3.9. Livestock and poultry manure generation rates

Animal	Manure Production Rate (dry kg/head-day)
Cattle	4.64
Hogs and pigs	0.56
Sheep and lambs	0.76
Chickens	0.025
Commercial broilers	0.040
Turkeys	0.101

Source: Stanford Research Institute, 1976, *An Evaluation of the Use of Agricultural Residues as Energy Feedstock*, Vol. 1 (Washington, D.C.: National Science Foundation Report), NSF/RANN/SE/GI/18615/FR/76/3.

Table 3.10. Annual biomass production potential in the United States

	Biomass Production (10^6 dry ton)	Heating Value (GJ/dry ton)	Energy Potential (10^9 GJ)
Agricultural residues	203	17.0	3.5
Forestry residues	167	19.0	3.2
Municipal solid waste	111	12.7	1.4
Excess cropland (Year 2012)	680–1,361	19.0	13–26
Potential cropland	405–1,087	19.0	7.7–21
Forest land	262	20.0	5.2
Total	1,828–3,190	—	34–60

Source: Production data from C. E. Wyman and B. J. Goodman, 1993, "Biotechnology for Production of Fuels, Chemicals, and Materials from Biomass," *Appl. Biochem. & Biotech.* 39/40, 41-59.

order of 3 quads of energy per year (a quad, the customary unit for measuring national energy consumption, is one quadrillion British thermal units, or $1.054 \times$ x 10^9 GJ). Municipal solid waste can supply half of that amount, about 1.5 quads. Since energy consumption in the United States is about 97 quads per year, biorenewable resources in the form of residues and wastes can supply nearly 8% of the nation's energy demand.

The greatest potential for biorenewable resources is lignocellulosic crops. Estimating the resource base for dedicated energy crops is difficult because of the large number of factors influencing decisions about how much land to devote to these crops. Clearly, competition for arable land in the production of food and feed will ultimately set a cap on the amount of land a nation can use to grow crops for fuels, chemicals, materials, and energy. Concerns that forests or ecologically fragile lands would be converted to crop production must be considered in making estimates of land availability, as well. Uncertainties about climate and productivity of dedicated energy crops also complicate the analysis. Table 3.10 includes three additional entries for lands that might be used in the future for growing biorenewable resources: *excess cropland, potential cropland, and forestland. Excess cropland* is land currently in crop production in the United States that is expected to be idled by 2012 due to improved agricultural productivity worldwide. *Potential cropland* is land that is not currently used for growing crops (such as pasture land) but could be converted to growing biorenewable resources. *Forestland* is low productivity forest that could be converted to growing short-rotation woody biomass. The first two of these three categories show a range of production capacity because of the uncertainties previously described.

These three categories, excess cropland, potential cropland, and forestland, represent an additional 26–52 quads of energy for the United States. Converted forest lands, planted to short-rotation woody crops, represent 10% of the total biorenewable resource of the United States while potential cropland and excess cropland represent 30% and 40%, respectively, of the total resource.

According to this estimate, biorenewable resources could provide 35–62% of current U.S. energy demand. Combined with energy efficiency improvements, the fraction of U.S. energy demand from biorenewable resources could be even more significant. A 20% reduction in energy demand would result in U.S. annual energy consumption of only 77 quads, a level that existed as recently as 1985. Under this scenario, biorenewable resources could replace 44–78% of U.S. energy demand.

Calculating the potential for biorenewable resources to replace imported petroleum resources is more difficult because, as will become apparent in subsequent chapters, a kilogram of cellulose is not equivalent to a kilogram of hydrocarbon in generating the fuels and chemicals to which we are accustomed. However, on an energy basis, biorenewable resources have the potential for replacing all 38 quads of petroleum consumed in the United States in 2002.

Table 3.11. Annual livestock and poultry manure generation in the United States

Animal	Population (10^6)	Production (10^6 dry ton)	Heating Value (GJ/dry ton)	Energy Potential (10^9 GJ)
Cattle	103.3	174.9	15.73	2.751
Hogs and pigs	59.6	12.3	16.99	0.209
Sheep and lambs	8.9	2.5	17.82	0.0446
Chickens	377.5	3.5	13.53	0.0474
Commercial broilers	7,018	103.2	13.53	1.396
Turkeys	289	10.7	13.49	0.144

Sources: Population data from U.S. Department of Agriculture, 1995, "Agricultural statistics 1995" (Washington, D.C.: National Agricultural Statistics Service); Manure production rates calculated from Table 3.10; Heating values obtained from Stanford Research Institute, 1976, *An Evaluation of the Use of Agricultural Residues as Energy Feedstock*, Vol. 1 (Washington, D.C.: National Science Foundation Report), NSF/RANN/SE/GI/ 18615/FR/76/3.

Note: GJ = gigajoules.

Table 3.11 estimates the total biorenewable resource in the United States in the form of livestock and poultry manure. The total amount, 4.6 quads, is slightly more than the estimate for agricultural residues. However, this material is often highly diluted in water and sometimes difficult to recover, reducing its attractiveness as a chemical and energy resource. Changes in the way that manure is collected and stored would be required in some instances.

Further Reading

Defining the Resource
Wright, L. and L. Hohenstein, eds. 1994. "Dedicated Feedstock Supply Systems: Their Current Status in the USA." *Biomass and Bioenergy*, 6(3).

Properties of Biomass
Alén, R. 2000. "Structure and Chemical Composition of Wood." *Forest Products Chemistry*. Per Stenius, Ed. Helsinki:Fapet Oy.

Rowell, R. M., R. A. Young and J. R. Rowell. 1997. *Paper and Composites from Agro-Based Resources*, CRC Press.

Rowell, R. M., T. P. Schultz and R. Narayan, eds. 1992. *Emerging Technologies for Material and Chemicals from Biomass*, ACS Symp. Series 476. Washington, D.C.: American Chemical Society.

Yields of Biomass
Heid, W. J., Jr. 1984. "Turning Great Plains Crop Residues and Other Products into Energy." *Agricultural Economic Report No. 523*. Washington, D.C.: U.S. Dept. of Agriculture.

Wayman, M. and S. Parekh. 1990. *Biotechnology of Biomass Conversion: Fuels and Chemicals from Renewable Resources*. Philadelphia: Open University Press.

Size of Resource Base

Wyman, C. E. and B. J. Goodman. 1993. "Biotechnology for Production of Fuels, Chemicals, and Materials from Biomass." *Appl. Biochem. & Biotech.* 39/40: 41–59.

Problems

3.1 Define dedicated energy crop. What are the two broad categories of dedicated energy crops?

3.2 Provide a definition for each of the following categories of woody crops and give their relative advantages as biorenewable resources.
    (a) Hardwoods
    (b) Softwoods

3.3 Provide descriptions of each of the following categories of herbaceous crops, and give their relative advantages as biorenewable resources.
    (a) Thick-stemmed grasses
    (b) Thin-stemmed grasses

3.4 The large volume of biomass that must be transported to a central processing facility may severely limit the practical size of such facilities. Calculate the maximum capacity (liters per year) of a fuel ethanol plant that uses rice hulls for feedstock if no more than 15 trucks per hour can enter the facility. Assume that each truck can carry 20 cubic meters of rice hulls and that the conversion of hulls to ethanol is 300 L/ton.

3.5 The average corn yield in a particular agricultural region is 150 bushels per acre of land. If the conversion of corn stover to ethanol is 315 L/ton, how many acres are required to operate a 5000 barrel per day plant (assume that the plant operates 350 days per year).

3.6 Overall energy use in the United States is expected to grow from 94 quadrillion Btu's in 1996 to 118.6 quadrillion Btu's in 2020. If this energy growth were completely provided by biomass, approximately how many acres of cropland would be required to provide this energy? How does this requirement compare to the biorenewable resource base in the United States?

3.7 In contemplating the use of biorenewable resources for the production of organic chemicals, the cost of carbon from various resources is often calculated. Estimate this cost for the following resources:

(a) Coal selling at $1/MMBtu
(b) Natural gas selling at $2.50/MMBtu
(c) Petroleum selling at $25/barrel
(d) Switchgrass selling at $40/ton
(e) Corn grain selling at $2.50/bushel

Production of Biorenewable Resources

4.1 Introduction

Traditionally, agriculture has been considered to be the cultivation of plants for food and animal feed, whereas silviculture is defined as the maintenance of forests, usually on time scales of decades, either for commercial, recreational, or ecological purposes. Neither term fully encompasses the activity of cultivating plants as dedicated feedstock supply for production of fuels, chemicals, materials, and energy. Generally, though, the production of woody or herbaceous material to be harvested annually or on a short rotation of 3–10 years more closely resembles agriculture than silviculture. Production of herbaceous crops and woody crops are treated in separate sections of this chapter because of the different time scales associated with their cultivation and the different approaches to their mechanized harvest. This chapter also considers the storage of biorenewable resources and the application of biotechnology to the production of transgenic crops.

4.2 Herbaceous Crops

As described in Chapter 3, *herbaceous crops* are plants that have little or no woody tissue and the aboveground growth usually lives for only a single growing season. These include *annuals*, which die at the end of a growing season and must be replanted from seed, and *perennials*, which die back each year in temperate climates but reestablish themselves each spring from rootstock.

Broadly defined, herbaceous crops include grasses, such as switchgrass, sugarcane, and cereals such as corn and wheat; pulses, which are leguminous plants such as soybeans, lentils, and alfalfa; and tubers, including potatoes, taro, and Jerusalem artichokes. All have potential application as part of dedicated feedstock supply systems. However, most development has centered on grasses, which are also the focus of this chapter.

4.2.1 Site Preparation

Preparation of the growth zone in soil for plant development is known as *tillage*. Typically this tillage zone extends 10–90 cm into the soil. The depth of organic matter in the soil defines the tillable zone. Tillage is required in virgin soils to clear it of plant matter for agricultural use. It also may be performed as part of the process of transferring seeds or seedlings into the soil in a manner conducive to healthy growth of the plants. Ideally, the seed bed is stratified, consisting of a base of coarsely loosened soil, a root development zone of porosity conducive to capillary movement of water, a seed bed zone of fine soil to surround the seeds, and an upper zone of small clods to protect the seeds, as illustrated in Fig. 4.1. Tillage is also used to control weeds and other pests living in the soil, although chemical means of pest control have largely supplanted mechanical methods of control.

Texture is a way to classify different sizes of particles making up soil. These classifications include *clay* (particles smaller than 0.002 mm), *silt* (particles between 0.002 mm and 0.05 mm), *sand* (particles between 0.05 mm and 2 mm), and *gravel and cobbles* (particles larger than 2 mm). The latter particles are undesirable as they interfere with tillage and retain little organic matter.

Soils are classified according to the distribution of the first three size classes (clay, silt, and sand). Soil classifications are represented by the texture triangle shown in Fig. 4.2. Soil type greatly influences its workability and suitability for agriculture.

Sands and sand soils, also known as light soils and shown on the lower left-hand corner of the texture triangle, are easy to work at all moisture levels. However, water storage capacity is low while infiltration rate and water conductivity are high.

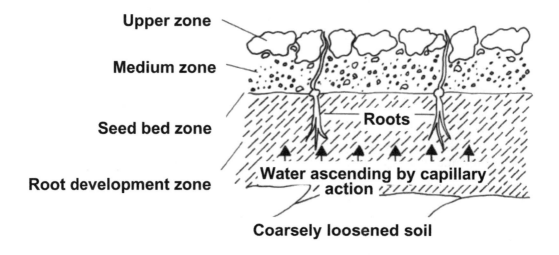

FIG. 4.1. Stratified seed bed of properly tilled soil (Source: B. A. Stout and B. Cheze, Eds., 1999, *CIGR Handbook of Agriculture Engineering, Vol. III: Plant Production Engineering* (St. Joseph, MI: American Society of Agricultural Engineers).

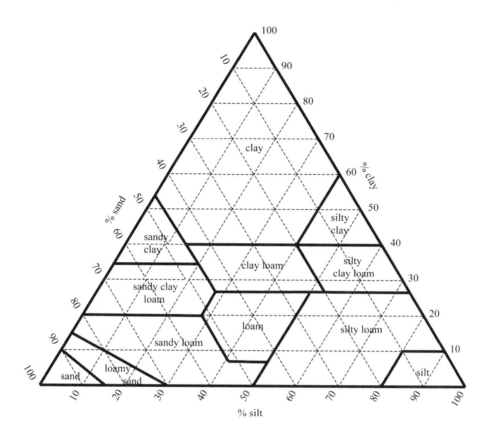

FIG. 4.2. Soil classification triangle (Source: B. A. Stout and B. Cheze, Eds., 1999, *CIGR Handbook of Agriculture Engineering, Vol. III: Plant Production Engineering* (St. Joseph, MI: American Society of Agricultural Engineers).

This is an advantage for irrigated soils, as salt does not accumulate. Crop yields for sand soils are lower than other types of soils because organic matter and nutrients are not readily retained.

Silty and loamy soils, also known as medium soils and shown on the lower right hand corner of the texture triangle, are the best agricultural soils. They readily retain water and have porosity suitable for good aeration. They readily retain organic matter and minerals required for good plant growth.

Clay soils, also known as heavy soils and shown on the upper corner of the texture triangle, are difficult to cultivate. At low moisture content they become hard, rendering tillage very difficult. At high moisture content they become plastic, which makes crumbling almost impossible and requires very high draft forces for tillage implements. Porosity is high, but the pores are so fine that aeration is poor and the contained water is inaccessible to plants.

One of the main objectives of tillage is achieving soil porosity that is conducive to both soil aeration and water accessibility to plants. An optimum tilth is usually obtained when crumbs are smaller than 50 mm and a sufficient number of clods exist to create porosity and protect the land against erosion. In the absence of clods, crumbs are readily broken up by wind and water into silty particles that form surface encrustations upon exposure to water. These encrustations interfere with plant growth. In humid climates, looser soil conditioning helps prevent encrustation and improves aeration. In dry climates denser soil conditioning improves water retention.

Figure 4.3 illustrates three types of tillage systems: *conventional tillage, reduced tillage,* and *no-till.* The practice of conventional tillage, characterized by high intensity of soil engagement and inversion of the soil, has rapidly declined in the last fifty years because of concerns about soil erosion. *Reduced tillage* encompasses a wide variety of cultivating practices that have in common less frequent and less intense tillage of the soil compared to conventional tillage. *No-till,* made possible by the development of chemical herbicides, entails very little disturbance of the soil in planting a crop.

Conventional tillage consists of primary tillage followed by several operations of secondary tillage. In heavy soils, primary tillage is accomplished with a moldboard

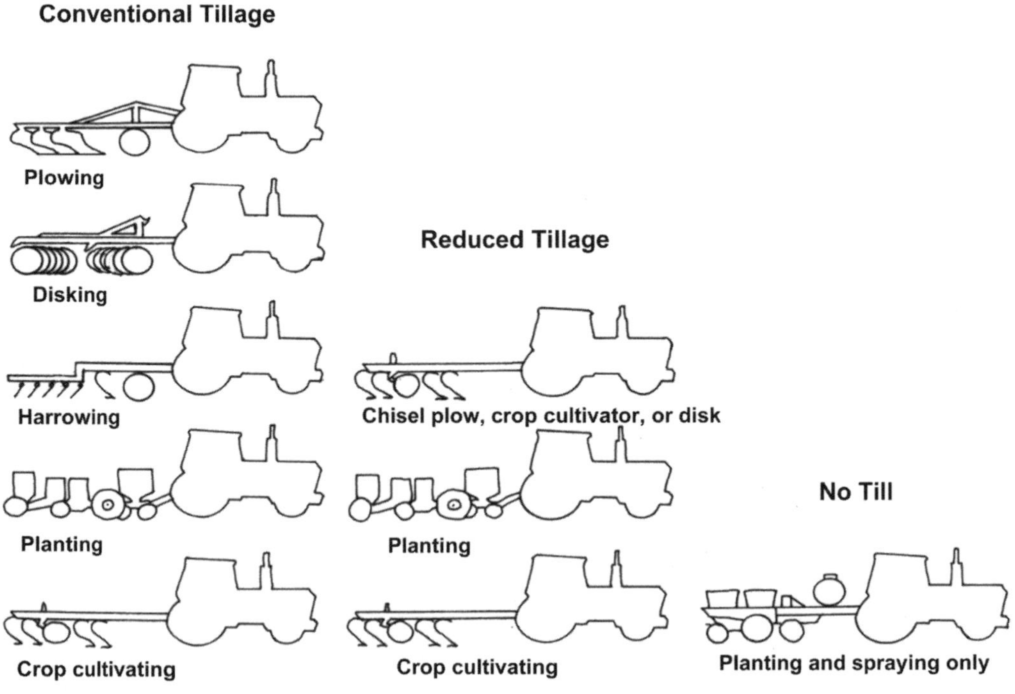

FIG. 4.3. Tillage methods.

plow. The soil is cut horizontally by a sharp steel element known as a share and lifted and turned over by a large planar surface known as the moldboard. In arid climates a disk plow is employed, which is less susceptible to obstacles and rough conditions. However, the disk plow is less effective at inverting and crumbling high-organic soils than is the moldboard plow. Primary tillage buries both weeds and plant residue from the previous year's crop. It also produces relatively large soil aggregates; thus, it must be followed by secondary tillage to create a fine seedbed. One of the most frequently used secondary-tillage implements is the disk harrow, commonly called a "disk," which consists of several sets of disks on a horizontal axis, angled into the direction of travel. Dragged over a plowed field, it crumbles, loosens, mixes, and levels the soil into a seedbed. A spiked-tooth harrow, resembling a garden rake in its action, is another implement of secondary tillage.

Reduced tillage is designed to reduce the number of steps in cultivating the soil, with corresponding savings in labor and energy inputs. Some forms of reduced tillage also constitute *conservation tillage*, which is defined as tillage practices that retain at least 30% of plant residue from the previous year's crop on the surface of the soil. This residue discourages movement of soil by wind and water erosion. Although several different methods of reduced tillage are practiced in different regions of the world, they generally do not employ moldboard or disk plows.

One system of reduced tillage is based on one or two passes of a disk harrow over a field followed by seeding. The greater the disk angle, the greater the degree of soil disturbance and the greater the amount of crop residue that is buried. Generally, not enough residue remains on the soil surface for this method to qualify as conservation tillage.

Another system employs one or two passes of a chisel plow, illustrated in Fig. 4.4a. Disk-shaped blades, known as coulters, at the front of the implement cut crop residue, preventing it from plugging the rows of chisels at the back of the implement.

(a)

(b)

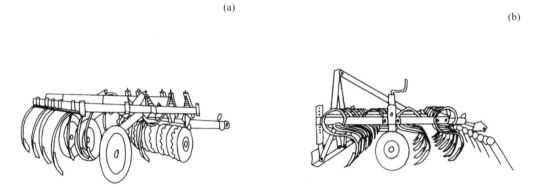

FIG. 4.4.   Reduced tillage implements: (a) Stubble mulch chisel plow, (b) Spring-tined cultivator.

The chisel plow cuts the soil to depths of 15–20 cm without burying the entire crop residue. Typically, more than 30% of the residue remains on the soil surface, thus qualifying as conservation tillage. Alternatively, a field cultivator, illustrated in Fig. 4.4b, is used either in place of the chisel plow or for the second pass over the field after the chisel plow. The field cultivator does not cut the soil as deeply as a chisel plow, and so helps conserve soil moisture, but it is not as effective in reducing weeds.

Zero or no tillage plants seeds directly into essentially unprepared soil. A typical machine for no till is illustrated in Fig. 4.5. This machine cuts a line in the soil for insertion of seeds, but does not otherwise disturb the soil. Its success is premised on the use of herbicides to kill weeds before planting.

4.2.2 Seeding and Planting

The final step in planting is insertion of seeds, seedlings, or vegetative propagations into the prepared soil. The term "planter" is used to describe machines that insert seeds into the soil as well as to machines designed for planting seedlings or vegetative propagations.

A simple system of metering a stream of seeds from a hopper through a feeder tube into a furrow is known as *bulk drilling* and is commonly employed for closely spaced crops such as small grains and grasses. In contrast, widely spaced crops, such as corn and soybeans, are planted using *precision drilling*, which aims at equidistant spacing of seeds. A modern planter designed to separate seeds for precision drilling is illustrated in Fig. 4.6. Drilling depth depends on seed size and water content of the soil: the larger the seeds and the dryer the soil, the deeper the seeds are planted to ensure emergence.

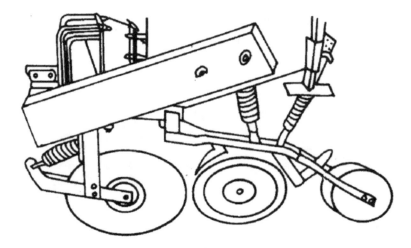

FIG. 4.5.   Zero till planter.

FIG. 4.6. Precision seed drill (Adapted from drawing courtesy of Deere and Company).

The planting of parts of plants, such as potato tubers, and whole plants, such as trees, has also been mechanized. As might be expected, the machines for vegetative planting are more complicated than those used for seed planting and are designed for specific kinds of plants.

4.2.3 Fertilizer Application

Natural soil fertility declines with crop production because a large fraction of the biomass is removed from the land, taking with it essential nutrients. *Fertilizer* is natural, or manufactured, materials containing high levels of nutrients essential to plant growth. The major nutrients are nitrogen (N), in the form of nitrate, ammonia, or urea; phosphorous (P), analyzed in terms of P_2O_5; potassium (K), analyzed in terms of K_2O; and calcium (Ca), analyzed in terms of CaO. Although fertilizer requirements vary with soil type and climate, estimated application rates for several biomass crops are listed in Table 4.1.

Fertilizer spreaders are designed for uniform distribution of powdered, granular, or liquid fertilizer. This is critical for optimum performance, as either excessive or inadequate application of the chemical will reduce crop yield.

A number of distributor types have been developed for powdered and granular fertilizer, including spinning disk spreaders, oscillating spout spreaders, pneumatic

Table 4.1. Estimated fertilizer requirements for selected crops

Biomass	Required Mass Per Unit Weight of Whole Dry Plant (kg/dry t)			
	N	P_2O_5	K_2O	CaO
Alfalfa	0	12.3	34.0	20.7
Corn	11.8	5.7	10.0	0
Kenaf	13.9	5.0	10.0	16.1
Napier grass	9.6	9.3	15.8	8.5
Slash pine (5 year)	3.8	0.9	1.6	2.3
Potato	16.8	5.3	28.3	0
Sugar beet	18.0	5.4	31.2	6.1
Sycamore	7.3	2.8	4.7	0
Wheat	12.9	5.3	8.4	0

Source: W. L. Roller, et al., October 1975, "Grown Organic Matter as a Fuel Raw Material Resource," NASA Report CR-2608 (Washington, D.C.: National Aeronautics and Space Administration).

spreaders, and auger spreaders. The *spinning disk spreaders* consist of a conical hopper feeding a spinning disk that is fitted with several radially oriented vanes. Fertilizer falling onto the disks is directed outward along the vanes at velocities of 15–50 m/s yielding a large broadcast area for the fertilizer. *Oscillating spout spreaders* consist of a mechanically agitated hopper connected to a spout that vibrates in the horizontal plane. Fertilizer granules are accelerated by the centrifugal forces in the spout, bouncing along the wall a couple of times before being discharged. *Pneumatic spreaders* entrain granules in a fast-flowing air stream. The air transports the fertilizer through long tubes supported by booms. Deflector plates at the end of the tubes spread the suspension into fans, which provide uniform coverage over the field. *Auger spreaders* deliver fertilizer from a hopper to spreading auger booms by means of scrapper floor chains or rubber auger belts. They are preferred for applying high rates of dusty soil ameliorants, such as potash or ground lime.

Liquid fertilizers are of three types: anhydrous ammonia, solution fertilizers, and suspension fertilizers. Although technically more difficult to apply, they have several advantages compared to granular fertilizers including more effective application and the ability to combine fertilizer application with pest control. Because of its high vapor pressure, *anhydrous ammonia* should be injected directly into the soil to prevent its dispersion into the atmosphere. *Solution fertilizers* consist of soluble fertilizers, such as ammonia, urea, or ammonium nitrate, dissolved in water. *Suspension fertilizers* employ a gelling-type of clay as an emulsifier to keep finely divided particles of fertilizer in suspension. This is an effective method for applying potassium fertilizer. Both solution and suspension fertilizers can be applied with field sprayers. Directing liquid over spinning fans forms coarse sprays suitable for application to the soil. Liquid nozzles form fine sprays suitable for direct application to plant leaves.

4.2.4 Pest Control

During the growing season, pest control may be required. *Pests* are anything that impede or compete with the desired crop and may include weeds, insects, or disease. Control includes mechanical, chemical, and biological methods as well as combinations of these methods, known as *integrated pest management.*

Mechanical weeding removes competing plants from the soil. A number of machines have been developed to support weeding in mechanized farming. The *hoeing machine* consists of tines or rotary hoes that uproot weeds between rows to a soil depth of 5 cm. The treatment is effective on dry, compact soil and is usually combined with banding application of herbicides. *Blade cultivators* cut weed roots and rhizomes to a depth of 10 cm by means of horizontal blades mounted on shanks. *Brushing machines,* effective during early weed emergence, are sets of rotating brushes that uproot weeds between the rows. *Chain harrows* consist of sets of narrow, flexible, vertical steel spikes that scratch the soil and root out post-emergent weeds from pre-emergent crop or shallow-rooted weeds from deeper-rooted crops. The harrow is particularly suited to non-row crops.

Chemical pest control was extensively developed after World War II and is currently the dominant method of controlling weeds, insects, and diseases in crops. Chemical control is classified as either contact or systemic. *Contact methods* are dependent on local action of the chemical at the point of application, whereas *systemic methods* are based on absorption and transport of chemicals through the plant to the point of action. The active ingredients of chemical pest control are classified as herbicides, which act against weeds; insecticides, which act against insects; or fungicides, which act against fungi. Dry applications are used where water is scarce or for certain compounds for which powders are more effectively deployed than liquid sprays. Spray applications can be very effective under appropriate meteorological conditions. For example, wind speeds should be below 3 m/s to avoid spray drift, temperature above 10°C promotes leaf absorption, and slight rain or dew improves efficacy of chemical deposition on plant surfaces. A variety of sprayers have been developed including broadcast sprayers, directed or banded sprayers, and in-furrow sprayers.

Biological pest control is of growing interest as a means of reducing the sometimes-unfavorable environmental impact of chemical pest control. In this method, fields are deliberately infested with predators and parasites of the pest. The eggs, larvae, or adults of pest predators are mixed with inactive material such as sawdust or hulls and distributed manually by workers walking through the fields to be treated. Because of the multiplier effect of biological pest control (one predator can often consume multiple pests), application rates of the mixtures are low, usually in the range of 5–30 L/ha.

4.2.5 Harvesting

Harvesting is the process of gathering a mature crop from the field. Specific machines have been developed for handling different types of crops: cereals, forage, canes, fruits, nuts, and vegetables. Some of these machines are also able to separate the desired agricultural product from residual plant material (for example, separating corn kernels from cob, husk, leaves, and stalk). In the case of grain crops, this separation process is known as *threshing*. A machine that combines both harvesting and threshing of grain crops (cereals) is known as a *combine harvester*, or simply a combine. Since grains and seeds, forage crops, and canes are the agricultural crops of particular interest as feedstocks for biobased products, only harvesting methods appropriate to these three crops will be discussed.

4.2.5.1 Grain Harvesting

Combine harvesters can harvest a wide variety of grains and seeds, ranging from mustard seeds to broad beans. Combined losses of grain from cutting, threshing, and cleaning can be as low as 1–3%. Mechanical methods have increased productivity of grain harvesting from as little as 10 kg/man-hour in 1800 to 60,000 kg/man-hour today.

The major subsystems of a combine harvester include the gathering and cutting unit (head), the threshing unit, and the cleaning unit. Different kinds of gathering and cutting units are used for different kinds of crops. A head for beans or small grains, illustrated in Fig. 4.7, cuts and gathers the stalks and pods with a cutter bar and rotating reel. A head for corn strips the cob from the stalk, leaving the majority of the plant in the field while sending the cob into the combine for further processing.

The crop is conveyed to the threshing unit, which detaches the grain from the ears or pods by a combination of impact and rubbing. Two types of threshing units are available: a cylinder/walker combination and a rotary feeder/separator. The cylinder/ walker combination passes the cobs or pods between a cylinder and concave, illustrated in Fig. 4.7, which releases grain to a grain auger while straw enters the straw walker. The straw walker agitates the straw, sifting out grain and chaff not separated in the thresher. The rotary feeder/separator uses a large rotating shaft mounted with tines that convey the crop through the combine and separate the grain from cobs or pods. The straw drops out the back of the machine and is left in a windrow for later baling, or is baled directly by a baling attachment or press, or is scattered over the ground by a fan-like straw spreader.

Grain, as well as chaff and fragments of straw shaken out of the straw, fall into a grain auger where they pass to the main cleaning device, called the cleaning shoe. The

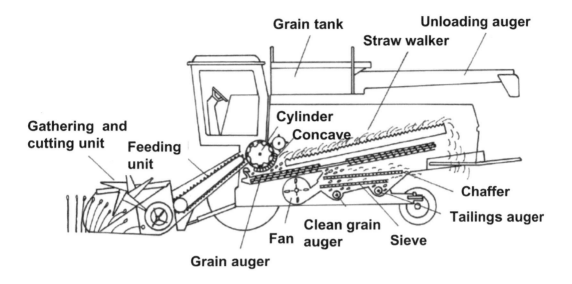

FIG. 4.7. Combine harvester (Adapted from drawings courtesy of Deere and Company).

cleaning shoe consists of one or two winnowing steps and two or three oscillating sieves. A winnowing step blows air across grain and material other than grain (MOG) as they fall from the end of the grain pan: chaff and pieces of straw are blown away while the grain falls onto a sieve (or a second grain pan, if there are two winnowing steps). The combined action of sieve oscillation and classification by air flow allows grain to penetrate the MOG layer and fall through the sieves to a grain auger which transports it to a grain tank. Material rejected by the top sieve is thrown from the back of the harvester while material rejected by the bottom sieve(s), which may still contain significant grain, is returned to the threshing unit by a tailings return system. Correct adjustment of the air blast, by adjusting fan speed, throttling the intake of the fan, or by altering the setting of baffles, is important in determining the degree of cleaning of the grain and the magnitude of the grain losses that occur.

4.2.5.2 Forage Harvesting

Forage crops are classified according to the method by which they are ultimately prepared for storage: hay or silage. The various crops used for these different preservation methods are listed in Table 4.2. *Hay* is preserved by storing at low moisture levels. Both grasses and legumes are harvested as hay with alfalfa being the most common hay crop. *Silage* is preserved at high moisture levels by fermenting it in the absence of oxygen thus producing organic acids that kill the microorganisms that cause spoilage. Corn is most commonly employed as a silage crop. *Haylage* is

Table 4.2. Forage crops classified according to preservation method

Preservation Method	Crop
Hay	Alfalfa, clover, sorghum, Birdsfoot trefoil, Reed canary grass, smooth bromegrass, Bermuda grass, wheat grass, Canada wild rye, timothy, Russian wild rye, native grasses, cereal grains
Silage or haylage	Corn, forage sorghum, Sudan grass, sorghum-sudan hybrids, oats, alfalfa, alfalfa-grass mixtures

Source: A. G. Cavalchini, 1999, "Forage Crops," in *CIGR Handbook of Agricultural Engineering, Vol. III: Plant Production Engineering*, B. A. Stout and B. Cheze, eds. (St. Joseph, MI: American Society of Agricultural Engineers).

a lower moisture version of silage, which further promotes preservation by limiting oxygen during storage. Distinct methods of harvesting and storage are employed for these two kinds of forage crops. Hay is usually harvested in operations requiring several passes over the field whereas silage or haylage crops are often collected in a single pass.

Since hay harvesting does not involve separation of plant parts, equipment much simpler than combined harvesting and threshing machinery can be employed. For haymaking this involves separate steps of cutting the crop (mowing), conditioning the cut crop to improve field drying (tedding), arranging the crop into long rows in the field (windrowing), and either baling the crop or loading the loose hay into wagons. The main parameters associated with the various operations of haymaking are summarized in Table 4.3.

Mowers include *rotary mowers*, which cut by means of impact forces, and *cutter bars*, which apply shearing forces to cut the plant material. *Rotary mowers* have high cutting speeds (10–12 km/h) and working widths of 1.5–3 m. However, they consume 20–25 kW-h/ha, which is about 40% higher than cutter bar mowers. *Cutter bar mowers* employ either an oscillating knife working in combination with a fixed finger bar or a mechanism of dual oscillating elements. The former cutter bar mechanism is the more robust of the two and is better suited for cutting crops close to the ground. Cutter bar mowers have mowing speeds of 8–12 km/h and working widths of 1.5–2.5 m. Energy consumption is 18–20 kW-h/ha.

Mowers often incorporate a mechanical treatment known as conditioning, which improves the uniformity of drying time for coarse stems and fine leaves. This not only slightly reduces field-drying times but, more importantly, prevents delicate leaves from overdrying and being lost during windrowing and baling. Conditioners can increase nutritive value of forage crops by as much as 15%. *Conditioners* consist of rollers for legumes, or flails for grasses, that slightly crush plant stems allowing more rapid release of moisture from this coarse plant part.

As part of haymaking, after the crop is cut and partially dried in the field it is turned and fluffed to promote additional drying. This process, known as *tedding*, was traditionally performed with a pitchfork. Modern machinery performs this

Table 4.3. Unit operations of haymaking

Operation, Machine	Speed (km/h)	Average Working Width (m) or Volume (m³)*	Capacity (ha/h,m) or (t/h)	Minimum Power (kW)	Energy Consumption (kW-h/ha) or (kWh/t)*	Time (man-h/ha) or (man-h/t)*
Mowing:						
Walking mower (finger bar)	2–3	1.2–1.4	0.2–0.25	8	18–20	4–5
Finger bar mower	5–7	1.5–2.5	0.4–0.5	15	18–20	0.8–1.6
Double knife cutter bar	6–9	1.5–2.5	0.5–0.7	15	18–20	0.6–1.2
Rotary disc mower driven from below	9–10	1.5–3.0	0.7–0.8	25	20–25	0.5–1.0
Rotary drum mower from the top	10–12	1.5–2.5	0.8–0.9	30	20–25	0.6–0.9
Tedding:						
Rotary tedder	10–14	2–6	1.5–6	20	10–15	0.2–0.7
Windrowing:						
Rotary rake	7–9	3–6	0.3–0.4	25–30	18–20	0.6–1.5
Parallel bar rake	6–7	2–3	0.25–0.35	15–20	15–18	1.0–2.0
Finger wheel rake	6–8	2–5	0.25–0.35	15–20	15–18	0.7–2.0
Mowing + Windrowing:						
Rotary drum mower driven from the top	10–12	1.5–2.0	0.8–0.9	30	20–25	0.6–0.9
Loading:						
Forage self-loading wagon	4–6	1.2–2.0	0.6–1.2	30	15–20	0.8–1.5
		15–30*	15–20*			
Baling:						
Conventional baler	4–6	1.0–1.8	2–8	25–45	1.8–2*	0.13–0.5*
Round baler	5–7	1.5–2.0	2–8	40–60	2.0–2.2*	0.13–0.5*
Big baler	6–8	1.8–2.2	10–15	70–100	1.9–2.1*	0.07–0.1*
Stack-wagon	4–6	1.6–2.0	10–15	40–70	1.5–1.8*	0.007–0.1*

Source: A. G. Cavalchini, 1999, "Forage Crops," in *CIGR Handbook of Agricultural Engineering, Vol. III: Plant Production Engineering*, B. A. Stout and B. Cheze, eds. (St. Joseph, MI: American Society of Agricultural Engineers).

*Values represent alternative units: column 3, (m³); column, 6 (kWh/t); column 7, (man-h/t).

operation by means of flexible forks on rotors mounted on a vertical axis: the forage is lifted off the ground and pitched backwards.

Windrower rakes distribute the cut crop into long rows in the field as a prelude to baling or collecting the material. Modern side delivery rakes include parallel-bar rakes and finger wheel rakes, illustrated in Fig. 4.8, as well as rotary rakes. Although an apparently simple operation, the ability to efficiently rake cut forage without contaminating it with soil while operating on irregular terrain is difficult to achieve. Leaf loss is a strong function of moisture content of the crop, ranging up to 8% loss for moisture content below 30%. Large, self-propelled windrowers produce a loose, fluffy windrow, making tedding unnecessary. Machinery has also been developed that combines mowing, tedding, and windrowing in a single operation.

Early methods of haymaking produced piles of dry, loosely consolidated plant material known as haystacks. Mechanization allowed hay to be compressed into rectangular bales of 15–30 kg weight that were easy to handle and store, helping reduce the labor for forage harvesting from 40 man-hours/ha to 20 man-hours/ha. However, with continued mechanization of agriculture, this conventional bale proved to be a bottleneck to further productivity improvements in haymaking. Modern handling systems include development of big bales and round bales, weighing as much as 700 kg, and systems for collecting loose hay in self-loading wagons or stack wagons. Pelletizing or cubing of forage to high densities (approaching 450 kg/m^3) has been explored as a means of facilitating handling and storage. Although technically feasible, the approach has only limited commercial application because of high capital costs, as well as energy costs in the range of 25–30 kW-h/t.

Figure 4.9 illustrates a big baler. Compression is usually achieved in two phases: in a prechamber the forage is accumulated and partially pressed, followed by final compaction in the upper compression chamber. In the upper chamber a plunger

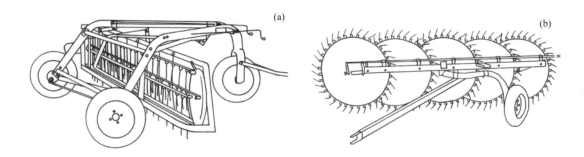

(a)   (b)

FIG. 4.8. Two types of hay rakes: (a) Parallel-bar rake, (b) Finger-wheel rake (Adapted from photographs courtesy of Deere and Company).

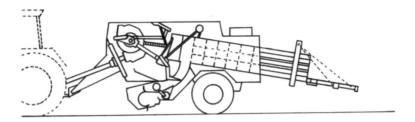

FIG. 4.9. Rectangular big baler (Source: B. A. Stout and B. Cheze, Eds., 1999, *CIGR Handbook of Agriculture Engineering, Vol. III: Plant Production Engineering* (St. Joseph, MI: American Society of Agricultural Engineers).

compresses the bales to densities in the range of 180–230 kg/m^3. Five to six binders are applied to hold the bale together. The bale length of 2.4 m is comparable to the length of the trucks designed to haul them. Offsetting the high collection rate of 15 t/h or higher are the high power requirements of the baling machine.

Figure 4.10 illustrates the two kinds of round balers: loose core and compact core. The *loose core baler* forms the bale in a chamber of fixed volume, which permits the bale to form around a relatively fluffy center with the periphery becoming more compact as the chamber fills. The disadvantage of reduced density is offset by greater air circulation within the core, which promotes additional drying

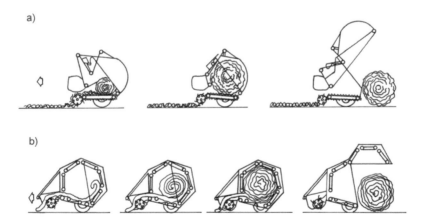

FIG. 4.10. Two kinds of round balers: (a) Variable chamber, core compacted, (b) Fixed chamber, loose core (Source: B. A. Stout and B. Cheze, Eds., 1999, *CIGR Handbook of Agriculture Engineering, Vol. III: Plant Production Engineering* (St. Joseph, MI: American Society of Agricultural Engineers).

without heat build-up after baling. The *compact core baler* employs a variable volume compression chamber, which provides constant compression during the accumulation of hay in the chamber.

In some parts of the world self-loading wagons, illustrated in Fig. 4.11, are used to harvest hay. They consist of a trailer with high sideboards and a cylinder type pickup combined with a forage chopping system. The chopping system produces lengths of hay that are 20–40 cm long for barn drying and 8–10 cm long for silage.

Stack wagons produce a fairly compact haystack of density between 100 and 150 kg/m^3 with a sloping top able to shed rainfall. The wagon, pulled by a tractor, has a flail type pickup consisting of horizontal rotors with flails or knives attached, which apply shearing forces on the forage. The forage is kicked up and transported pneumatically to a large, rectangular compression chamber constructed of sheet metal. The top surface is a movable canopy that allows the load to be compressed as the chamber is filled. Stack wagons are often employed to create winter feed stores to be left on the field.

The storage density and volume, bale weights, tractor power and energy requirements, and collection rates for these various options are summarized in Table 4.4.

Forage cereals, by virtue of their high soluble sugar content, must be ensiled for successful storage. Thus, harvesting occurs at relatively high moisture content: in the range of 32–38% depending on the kind of cereal. Forage harvesters can either

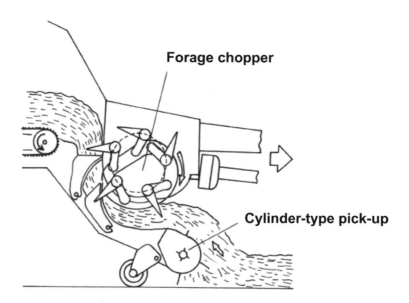

Forage chopper

Cylinder-type pick-up

FIG. 4.11. Self-loading wagon for hay harvest (Source: B. A. Stout and B. Cheze, Eds., 1999, *CIGR Handbook of Agriculture Engineering, Vol. III: Plant Production Engineering* (St. Joseph, MO: American Society of Agricultural Engineers).

Table 4.4. Characteristics of different hay collection methods

Collection Method	Hay Density (kg/m³)	Bale/Stack Dimension or Wagon Volume		Weight (kg)	Energy Requirement (kW-h/t)	Collection Rate (t/h)
Self-loading wagon	<100	10–20 m³		1000–3000	1–1.2	4–4.6
Stack wagon	100–150	Length (m)	2–5	1000–6000	1.5–1.8	10–15
		Width (m)	2–3			
		Height (m)	2.5			
Conventional bale	125–165	Length (m)	0.7–1.2	15–30	1.5–2.5	4–8
		Width (m)	0.40–0.5			
		Height (m)	0.35–0.45			
Big bale	180–230	Length (m)	1.0–2.4	200–700	2–2.5	10–20
		Width (m)	0.8–1.2			
		Height (m)	0.5–1.0			
Round bale	130–205	Length (m)	1.0–1.8	200–700	2–2.5	4–12
		Diameter (m)	1.0–1.8			

Source: A. G. Cavalchini, "Forage Crops," in *CIGR Handbook of Agricultural Engineering, Vol. III: Plant Production Engineering,* B. A. Stout and B. Cheze, eds. (St. Joseph, MI: Society of Agricultural Engineers).

collect windrowed material or cut a standing crop. Forage heads include a windrow pickup head for haylage, a mower bar for winter cereals or standing grasses, and a row crop head for corn silage. The forage is then fed to cutterheads, illustrated in Fig. 4.12, that chop the forage to suitable size, typically 6–9 mm for maize and winter cereals and 20–30 mm for grasses.

Machinery is being developed for the simultaneous harvest of corn grain and stover. Conventional combine harvesters collect the grain and leave stalks, leaves, husks, and cobs, collectively known as *stover,* in the field. If stover is to be harvested, a second pass over the field is required. A more economical approach to the collection of clean stover is to pick it up at the time the grain is harvested. Gathering heads have been adapted to combine harvesters that process the whole plant, sending the stover to a storage bin separate from that for grain. These systems show promise in producing stover free of dirt and containing essentially all the cobs. The amount of residues left on the field can be adjusted according to local soil conservation needs.

4.2.5.3 Cane Harvesting

Sugarcane can be harvested either as *whole cane*, the traditional method of harvesting that can be done either manually or mechanically, or as *chopped cane*, which was developed for harvesting by power-driven equipment. Whole cane harvesting is practiced where labor is inexpensive or where a gradual transition to mechanization is desired to prevent disruptions in existing harvesting and processing infrastructure. It also produces a very clean crop, important in fiber production. *Chopped cane harvesting* reduces the multiple steps of traditional harvesting into a single machine. Early attempts to adapt machines to local climate and geography

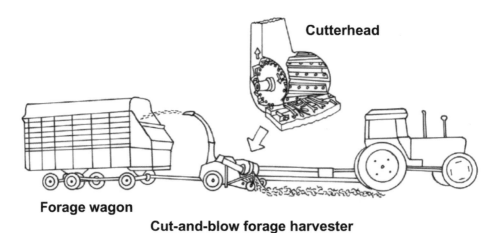

FIG. 4.12. Basic forage-harvester operation (Adapted from photographs courtesy of Deere and Company).

have been abandoned in favor of field and crop management practices that are con-ducive to mechanical harvesting, such as leveling ground and growing lodging-resistant varieties of sugarcane.

Manual cutting of whole cane involves three operations: cutting the cane at the base; trimming off the top; and laying the canes in windrows or heaps. Loading the windrowed or heaped canes onto wagons for transport completes the harvest. Mechanization of whole cane cutting imitates one or more operations of cutting, trimming, and windrowing. Some machines simply cut the cane and require fol-low-up by human laborers or machines, while other approaches completely mech-anize the cutting operations. Loading of either windrowed or heaped cane is performed discontinuously with front-mounted tractor loaders, self-propelled front-end loaders, swivel loaders, or power-loading trailers. Windrowed cane allows the option of continuous loading by means of wide chain elevators that pick up the whole canes, chop them into 40–50 cm lengths with circular saws, and drop them into a following wagon.

Chopper harvesters completely mechanize the harvesting operation, including the loading of cane into a trailer. The cane is cut at the base, chopped into 20–40 cm lengths, and loaded directly into a trailer. The cane can be chopped immedi-ately after cutting, which is known as *bottom chopping* (that is, near the ground), or after it is conveyed through the machine, which is known as *top chopping* (that is, elevated from the ground). The advantage of bottom chopping is reduced power requirement by the machine. The advantage of top chopping is cutting blades are protected from dulling by soil and rocks.

The method of transporting cane to sugar mills depends on the distance involved. Field tractors and trailers are frequently used for distances of less than 10 km. High capacity road trailers or tractor-trailers are used for longer distances. Occasionally, private rail systems are employed for large production systems. In any case, cut cane must be delivered to the mill and crushed within 25 hours if the field was burned before harvest or if cut green, within 48 hours to prevent sugar loss.

4.3 Woody Crops

The term *short-rotation woody crop* (SRWC) is used to describe woody biomass that is fast growing and suitable for use in dedicated feedstock supply systems. Desirable SRWC candidates display rapid juvenile growth, wide site adaptability, and pest and disease resistance. Promising tree species include poplar (*Populus* spp.), willow (*Salix* spp.), silver maple (*Acer saccharinum*), sweetgum (*Liquidambar styraciflua*), sycamore (*Platanus occidentalis*), black locust (*Robinia pseudoacacia*), and *Eucalyptus* spp. Trees of potential regional importance in the United States include alders (*Alnus*), mesquite (*Prosopis*), and the Chinese Tallow (*Sapium sebiferum*).

Hybrid poplar and eucalyptus are most promising for the United States because of high growth rates ranging between 20 and 43 Mg/ha/yr. In the United States, hybrid poplar has a wider range than *Eucalyptus*, which is limited to southern Florida, California, and Hawaii. Hybrid poplar is also attractive for the ease of propagating it from either stem cuttings or tissue culture.

4.3.1 Site Preparation

Short-rotation woody crops need to be established under conditions similar to those for herbaceous crops. Deep, well-drained fertile soils with adequate rainfall provide the best yields. Light-textured soils such as sandy loams or silty loams are preferred, but heavier-textured soils can be employed with sufficient drainage. Thus, suitable acreage ranges from good cropland to land that is marginal for conventional crop production because of poor drainage or erosion potential.

Although hybrid poplar may have the best overall potential for dedicated feedstock supply systems, other species are better adapted to certain regions or marginal lands. Sycamore and sweetgum are more suitable for southeastern United States where respiration rates are higher and drought more common. Silver maple is well adapted to bottomlands susceptible to flooding, while black locust, which can extract nitrogen from the atmosphere (nitrogen fixation), can be grown on nutrient-poor soils. Short-rotation woody crops are unsuitable for cutover upland forest sites because of generally thin soils, low fertility, and susceptibility to erosion.

Site preparation should begin 1 year before planting. Abandoned cropland or pasture typically requires brush removal, mowing, plowing, and application of broad-spectrum herbicide to eliminate competing plants. Conversion of row-crop fields to SRWC may only require a single spring plowing and row-marking operation. Conventional farm tractors and implements, such as moldboard plows, discs, and harrows, can be used for site preparation.

4.3.2 Seeding and Planting

Seeding is not directly employed to establish trees in dedicated feedstock supply systems. Instead, bareroot seedlings, containerized stocks, or cuttings are planted in the prepared site. Large-seeded species can be sown in beds where they sprout and grow into seedlings that are transplanted as bareroot seedlings. *Containerized stock* are trees grown in plastic or fiber pots, a system that generally produces trees with higher survival rates upon field planting than do the alternative propagation methods, but at a cost that is 25–250% higher than the alternatives. *Cuttings* are plants reproduced by *clonal propagation*, a process by which tissue taken from the root, stems, or leaf of a plant is grown into a whole plant genetically identical to the original plant. Clonal propagation is favored for dedicated feedstock supply systems because trees with the most desirable properties, such as disease resistance or rapid growth under specific local site conditions, can be selected and rapidly multiplied.

In some instances, clonal propagation requires culturing the vegetative tissue in special "cloning solutions" until roots are established, at which point the cutting can be planted to the field or a container. However, cuttings from many hardwood species can be planted directly into the ground. Most planting material is 25-cm-long hardwood cuttings with diameters of 1–2 cm. Species that sprout and grow readily include many *Eucalyptus* spp., *Populus* spp., and *Chlorophora excelsa*.

Spacing of plantings is critical as it affects establishment costs, optimal rotation ages, and tree size at harvest. Optimal spacing is a function of the tree species, soil quality, climate, and desired size at harvest. Narrow row spacing has the advantage of allowing tree crowns to shade out competing weeds within three or four years and reducing branching. Tree spacing ranges from 1 m $\times$ 1 m for willow plantations that are harvested every 2–3 years (10,000 trees per ha), to 4 m $\times$ 4 m for large saw logs (625 trees per ha). Planting densities of 1,500 trees/ha might be considered typical. Both hand planting, at rates of 4 ha/day/person, and machine planting, at rates of 16 ha/day for a 3-person crew, have been employed. Machine planting can be problematic in rocky soils, wet soils, or poorly prepared soils.

4.3.3 Fertilizer Application

Fertilization is required to maintain rapid growth rates as well as to maintain the fertility of the soil. For trees planted to fertile bottomlands, application of fertilizer may not be required until the second rotation of SRWC. Most sites will require nitrogen application by year three. Granular nitrogen fertilizer can be applied by either ground or aerial equipment. Ground application of granular nitrogen is done with cyclone-type applicators.

4.3.4 Pest Control

Good survival and growth of SRWC can be achieved compared to either herbaceous energy crops (HEC) or conventional crops if the site is thoroughly prepared and planted with fast-growing clones. Herbicide applications before and after planting are usually employed to reduce labor of hand and mechanical weed control. Rotary hoes can be used to till newly established plantations on light (sandy) soils to control emerging weeds until the trees are 30 cm tall.

A large number of animal and microbial pests must also be controlled. These include deer, elk, beavers, voles, and gophers in the first two years of establishment and insect and fungal species in even well established stands. The best strategy is to select species and clones that are best adapted to pests indigenous to the region. For example, hybrid poplar is a poor choice for eastern United States because of insect and fungal pests. In some circumstances, application of insecticides or fungicides may be required to control these pests.

4.3.5 Harvesting

Rapid juvenile growth and high carbon sequestration rates can be sustained only if the stands are regularly harvested. Typically, rotations are no longer than 8–10 years as growth rates peak after 4–6 years and quickly slow as a result of competition among the trees.

Harvest should occur during the dormant season, which allows a significant quantity of nutrients to be translocated to the roots or returned to the soil by leaf fall. This assures rapid growth of coppice the following spring. There is little data on long-term productivity of SRWC grown in coppice systems. Some economic analyses project that stands will be harvested only three times, representing a total time period of 15–30 years before replanting, because improved tree varieties will become available in that time.

Harvesting equipment for traditional logging operations was developed for broken terrain typical of upland forests as opposed to the relatively flat agricultural lands to be planted to short-rotation woody crops. Much of this equipment is adapted for coniferous (softwood) trees that are larger and less uniform than the deciduous (hardwood) trees of interest for dedicated feedstock supply systems. Thus, an evolution of harvesting equipment can be expected for woody crops as the industry develops.

Harvesting in the near term will employ several pieces of machinery familiar to the conventional logging industry. A *feller/buncher*, illustrated in Fig. 4.13, performs the initial operation of cutting the trees (felling) and stacking them (bunching). This self-propelled machine, which is either rubber-tired or tracked, has a felling head at the end of an articulated, extensible arm. The felling head consists of a grappling device and either a disc or chain saw. The human operator uses the grappling device to grasp the trunk of a single tree or possibly the trunks of up to six small trees. The saw severs the tree trunk from the stump, and the operator maneuvers the articulated arm to rotate the trunk to a horizontal position and lay it into a pile of felled trees.

A *grapple-equipped skidder*, illustrated in Fig. 4.14, transfers the pile of felled trees to a centralized "landing" for further processing. The grapple skidder is a rubber-tired, four-wheel drive vehicle with a dozer blade on the front and a maneuverable grappling device on the back. The skidder is backed into position next to a pile of trunks and the grapple used to pick the pile off the ground before transporting it to the landing.

At the landing, limbs and bark are removed mechanically, usually by a *flail*, which is a rotating drum fitted with lengths of steel chain that batter the tree trunks and break off limbs and shatter bark. The mixture of bark and limb wood is usually suitable only as fuel wood. The delimbed and debarked tree trunks are

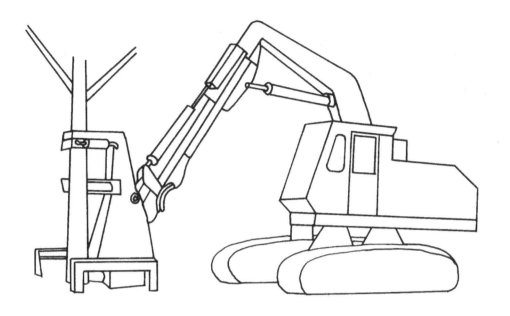

FIG. 4.13. Feller/buncher for cutting and piling woody crops.

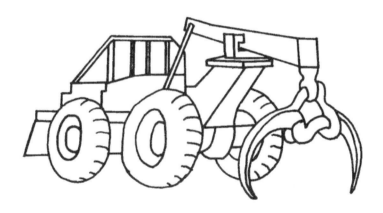

FIG. 4.14. Grapple skidder for moving piles of felled trees to a "landing" for further field processing.

then fed to a tub grinder, also located at the landing, which produces chipped pulpwood suitable as feedstock to be processed into chemicals or fibers. The combined operations of delimbing, debarking, and chipping are illustrated in Fig. 4.15.

Flail delimbing and debarking is the bottleneck in producing clean feedstock from woody crops. High production rates yield unacceptably high levels of bark

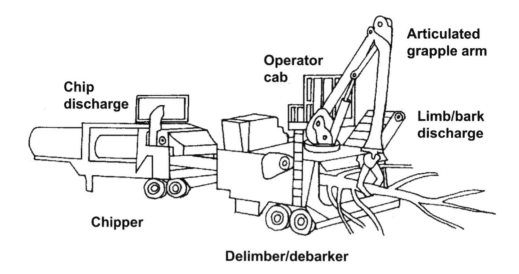

FIG. 4.15. Flail delimber/debarker/chipper operation.

mixed with the pulpwood, a level that must be less than 1% to be suitable. Furthermore, the process entails relatively high operating costs, including energy consumption (40–70 MJ per dry Mg of residues) and chain replacement.

Equipment optimized for harvest of short-rotation wood crops is still being developed. Some of this equipment resembles the mechanical harvesters used to cut and collect bundles of sugarcane stalks. Others are essentially heavy-duty forage harvesters like those used to chop corn stalks into silage, except with larger cutting heads. This development will probably proceed slowly until widespread commercial markets develop for the pulpwood.

4.4 Storage

Agricultural production is seasonal, with crops being harvested only once or just a few times per year. But processing to energy, fuels, chemicals, or materials is a year-round activity as manufacturing facilities are too expensive to remain idle most of the year. Most biomass undergoes substantial degradation in the course of a few months or even days if exposed to the elements after harvest. Accordingly, proper storage to preserve plant materials for periods of a year or even longer is critical to the successful development of a biobased products industry. Traditional methods of long-term storage involve drying, cooling, or ensiling a crop or immediately processing the crop to obtain a more stable intermediate product. Each method has its advantages and disadvantages, the choice depending on the nature of the crop and the kind of processing it will ultimately undergo.

4.4.1 Drying

Drying removes moisture from a crop to attain a moisture level low enough to impede the growth of microorganisms. Freshly cut biomass may have a moisture content of 60% or higher for woody biomass and 70–85% for herbaceous biomass. This moisture exists in two forms: free water within the pores of the plant material and bound water absorbed in the interior structure of the material. Successful preservation of plant material may require drying to as little as 10% moisture. Drying is a very energy-intensive process, theoretically requiring 2442 kJ of energy for every kilogram of moisture removed at 25°C. In practice, drying, which is often performed at temperatures slightly higher than 100°C, requires about 50% more thermal energy than this theoretical level due to the sensible heat of both the biomass and of the air used for drying. To dry a ton of biomass containing 50% moisture down to 10% moisture would require about 1.5 GJ of energy, representing about 18% of the energy content of the fresh biomass. Thus, field drying is employed whenever possible.

A crop left standing in a field beyond maturity will naturally dry, as is commonly done for such crops as corn, soybeans, and wheat. Green crops can also be cut and left lying in the fields to dry, as is commonly practiced in haymaking. The level of drying depends upon the climate, the kind of crop, and the structure of the plant part. For example, leaves dry more readily than stems, and plant parts protected by pods or husks will dry more slowly than exposed plant parts. In many regions of the world, grasses can be field dried to moisture levels consistent with long-term storage. Many grains, though, protected as they are by pods and husks, require additional drying in grain silos after they have been harvested, depending on local climatic conditions.

Barn drying of hay is employed in climates where field drying is not practical. After the forage is cut, it is allowed to field-dry down to a moisture content of 40–45%. The material is then collected with a self-loading wagon as previously described and stored in a drying barn, where either ambient air or air that has been heated 10–20°C above ambient is circulated through the hay to reduce moisture to around 15%. Barn drying is often employed in small to medium-sized dairy farms.

Baling, as previously described, is appropriate for hay and crop residues, such as corn stover. Conventional balers do not yield sufficient compaction for economic storage, especially for stalks. Large round or square bales are more appropriate for long-term storage. Large round bales can be stored outside for several months, but there can be some losses due to water damage and weathering. Bales can also be wrapped in polyethylene sheets, canvas, or nylon tarpaulins to help shed water during field storage. Ideally, bales are stacked in sheds to protect them from the weather. Handling of baled crops is very labor intensive, which makes this mode of storage unattractive in places where development has increased the cost of labor.

4.4.2 Cool Storage

Storing biomass in a cool, dry environment is also an effective method for discouraging microbial degradation. Farmers traditionally used root cellars, consisting of excavations into hillsides, as cool places to store tubers for periods longer than would otherwise be possible for these relatively high moisture crops. Today refrigerated rooms are used to store high-value crops, such as seed corn, but use of cool storage is not practical for commodity crops.

4.4.3 Ensiling

Ensiling was developed for humid climates where field drying is impractical, but it is finding increasing application in drier climates. Ensiling is attractive for automating the handling and storage of cereals, grasses and legumes, bagasse, and even corn stover, the latter of which is typically much drier than traditionally ensiled crops. Grasses and other graminaceous species are easier to ensile than legumes, the protein content of which promotes butyric fermentations associated with rotting of the forage.

The crop is ideally harvested at a moisture content of 40–50% and stored under anaerobic conditions to promote partial fermentation of sugars to organic acids, which act as a preservative and suspend further microbial degradation of the crop. Chemical additives, usually organic acids, water-soluble salts, or pulverized limestone, are sometimes added for pH control.

Storage systems include horizontal silos and vertical silos. *Horizontal silos* include bunkers, consisting of an above ground structure of concrete floors and walls; trenches dug into the ground, with either bare earth or concrete floors and walls; large polyethylene bags; and unprotected stacks for temporary storage. *Vertical silos* include conventional silos, constructed from metal, concrete, or tile, and oxygen-limiting units usually constructed of metal with an inner lining of fused glass. The technology is relatively easy to implement and requires no direct energy input.

Losses during ensiled storage are inevitable. If properly preserved, biomass losses can be kept to only 5–10%, depending upon the sugar content of the ensiled material (for example, *bagasse*, the fibrous stalks remaining after expressing juice from sugarcane, has a relatively high residual sugar content). Improper ensiling can result in losses as high as 30%.

4.5 Transgenic Crops

For thousands of years mankind has employed plant breeding to improve yields of sugars, starch, oil, and protein in crops, increase resistance to pests, and adapt plants to local climatic conditions. In conventional plant breeding, genes already existing within a species are brought together in new combinations by sexual crossing in an effort to express desirable characteristics. Statistical laws control the possible outcomes, but careful selection of offspring with desired traits makes the process deliberate. Recent advances in biotechnology broaden the scope of plant breeding by allowing selection of desirable traits across species, thus the name *transgenic crops*. Because genetic material is directly manipulated, the process is also less probabilistic, the outcomes known *a priori*.

4.5.1 Genetic Manipulation

The process of creating a transgenic plant, illustrated in Fig. 4.16, begins by identifying an organism, whether bacteria, plant, or animal, that contains the desired trait to be expressed in the new plant. The expression of this trait is mediated by various proteins that serve as catalysts for biochemical reactions or as building blocks for cellular components. The genetic roadmap for manufacturing these proteins is the organism's DNA (deoxyribonucleic acid), usually contained within the cell nucleus. Because all organisms employ similar mechanisms for transcribing and translating the information contained in DNA into proteins, the DNA from one organism, in principle, can be inserted and used in another organism.

A gene is a segment of DNA that controls assembly of a particular protein. Thus, after *extracting* DNA from the organism that expresses the desired trait, the next step is to isolate the gene that produces the protein associated with the trait. Although tremendous strides have been made in "mapping" DNA for various organisms, the process of identifying and locating genes for agriculturally important traits will limit the rate of introduction of transgenic crops for several years to come. Usually, it is not enough to know which stretch of DNA produces a particular protein since other genes may also interact with the biochemical pathway to be manipulated. However, once the gene has been identified, it can be *isolated* by the use of restriction enzymes, which recognize and cut phosphate bonds at specific locations along the backbone of the DNA molecule, and then be further *manipulated* with ligases, enzymes that join fragments of DNA. The isolated gene is inserted into bacteria where it is replicated, or *cloned*, in sufficient quantities for subsequent genetic manipulations.

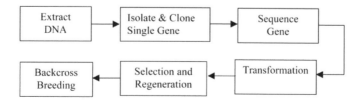

FＩＧ. 4.1 6.   Steps in producing a transgenic plant.

| Marker Gene | Promoter Sequence | Transgene | Termination Sequence |

FIG. 4.17. Components of a constructed transgene.

The isolated and cloned gene requires further genetic manipulation, called *sequencing*, before it can be successfully expressed as a transgene in a host plant. As shown in Fig. 4.17, this entails the addition of a selectable marker gene, a promoter sequence, and a termination sequence. The *selectable marker gene* is added as a means of identifying which plant cells have successfully integrated the transgene, since the insertion process has a relatively low probability of success. Selectable marker genes encode proteins that confer resistance to agents that are normally toxic to the plant being transformed. For example, the marker gene might protect against an antibiotic or herbicide. Thus, plants that successfully incorporated the transgene are identified by treatment with the toxic agent and selecting the surviving plants. The *promoter sequence* is the on/off switch that controls when and where the plant gene will be expressed. A "constitutive" promoter is commonly employed, which causes the gene to be expressed throughout the life cycle of the plant in most tissues. Other promoters respond to specific environmental cues, such as light. The *termination sequence* simply indicates to the cellular machinery that the end of the transgene has been reached.

Transformation is the process of inserting the transgene into the desired host plant. A wide variety of techniques have been developed to achieve transformation including plant viruses, chemically mediated DNA uptake, *Agrobacterium* co-cultivation, electroporation, microinjection, microprojectile bombardment, and electric discharge particle acceleration.

Certain kinds of plant viruses can introduce DNA into normal, healthy plants. Transformation may be as simple as rubbing the leaves of the target plant with viruses in which the transgene has been inserted: the virus systematically infects the cells of the plant, transferring the gene into the plant's genetic material. The most commonly employed viral vector is the cauliflower mosaic virus. Viral transformation is limited to plants it can naturally infect. It has little prospect for producing transgenic cereal crops.

Polyethylene glycol and polyvinyl alcohol, in combination with calcium ions and high pH, can produce chemically mediated uptake of DNA. The integrated gene was shown to be inheritable in subsequent generations of plants transformed in this manner. While transformation frequency is inferior to that of other techniques, in particular *Agrobacterium* co-cultivation, it may have application to graminaceous monocotyledons (species related to grasses), which are more difficult to transform by *Agrobacterium* co-cultivation (described below).

Agrobacterium co-cultivation involves the co-incubation of plant cells with *Agrobacterium tumefaciens*, a soil bacterium that naturally infects many dicotyledonous (broadleaf plants like soybeans and tomatoes) and gymnospermous (coniferous) plants. The DNA for this organism is contained in the bacterial chromosome as well as in a structure known as the Ti (tumor-inducing) plasmid. The Ti plasmid includes a section of DNA known as T-DNA, which is transferred to the plant cell during the infection process, and another section of DNA, known as vir (virulence) genes, which directs the infection process. To harness *A. tumefaciens* as a transgene vector, the tumor-inducing section of T-DNA is removed, while the T-DNA border regions and the vir genes are retained. The transgene is inserted between the T-DNA border regions, which is transferred to the plant cells during co-incubation and becomes integrated into the plant's chromosomes. The process has been widely employed because of the large number of cells that can be treated at a time and the uniform exposure that is achieved. However, many important monocotyledons, such as rice and corn, are not susceptible to transformation by this technique.

Electroporation achieves transformation by applying a high electrical potential to a mixture containing plant cells and the transgene. The process is hypothesized to induce cell membranes to a state of high-permeability, allowing transgenes to diffuse into the cells, although the exact mechanism is not fully understood. Electroporation has found application to monocotyledons and other plants that are recalcitrant to other transformation processes. The advantages of this approach include inexpensive instrumentation, reproducibility, and avoidance of toxic chemicals.

Microinjection uses microcapillaries and other microscopic devices to deliver DNA into individual plant cells. The potential for microinjection was first demonstrated in rapeseed, but the technique has not advanced as much as other transformation processes. As might be expected, the process requires considerable skill and specialized instrumentation, and only one cell receives DNA per injection.

Microprojectile bombardment accelerates micron-sized particles of tungsten or gold that have been coated with DNA to velocities sufficient for non-lethal cell penetration. This "biolistics" approach requires careful refinement to avoid excessive mortality of plant cells so treated. This transformation process has been extensively studied, especially for monocotyledonous species such as rice and corn. A closely related process, *electric discharge particle acceleration*, applies a high voltage electric discharge to a tiny water droplet, which rapidly vaporizes and serves as a propellant to DNA-coated spheres of gold. Uneven particle distribution and cell death by bombardment contribute to low transformation efficiency for these two processes.

From the above descriptions of transformation processes, it is evident that a relatively small fraction of attempts to insert a transgene into plant cells are successful. Accordingly, gene insertion must be followed by *selection* and *regeneration*, in which plant cells that have successfully incorporated the transgene are separated from the rest of the plant cells and grown into whole plants. The inclusion of a selectable marker gene in the constructed transgene makes this separation possible. The mixture of transgenic plant cells and normal plant cells from the transformation process are transferred to a growth medium containing the toxic agent for which the selectable marker gene affords immunity. Thus, only the plant cells that have successfully incorporated the transgene will survive, all others will perish from the toxic agent.

Those plant cells surviving the selection process are grown into whole plants under controlled environmental conditions in a series of growth media containing nutrients and hormones. These whole plants generally do not possess the qualities of modern cultivars demanded by producers and consumers. Therefore, the transgenic plant will be crossed with an improved variety. This initial cross to the improved variety must be followed by several cycles of repeated crosses to the improved parent, a process known as *backcross breeding*. The goal is to recover as much of the improved parent's genome as possible, with the addition of the transgene from the transformed parent.

The final step in the process is multi-location and multi-year evaluation trials in greenhouse and field environments to test the effects of the transgene and the overall performance of the cultivar. This phase also includes evaluation of environmental effects and food safety.

4.5.2 Biobased Products from Transgenic Crops

Genetic transformation of crops offers many opportunities for plant improvement, including resistance to herbicides, protection against insects and other pests, hardiness against frost, and the manufacture of valuable products in plants. This latter prospect has led to discussions about "plant factories" and "plant molecular farming." Biobased products that might be manufactured directly in plants to yield commercially recoverable quantities include antibodies; oligopeptides and proteins; sugar oligomers and polymers; phenolics and alkaloids; essential oils; monomers for biodegradable polymers; and enzymes for industry and therapeutic purposes.

An example of a particularly intriguing prospect is the growth of a class of industrial polymers called *polyhydroxyalkanoates (PHA)* in plants. This concept was based on the observation that the bacterium *Alvaligenes eutrophous* can directly synthesize poly-3-hydroxybutyrate (PHB), which accumulates as polymeric inclusions in cell bodies, from glucose. However, the relatively high cost of glucose and other suitable carbon sources makes the economics of PHA from bacteria unattractive compared to polymers derived from fossil resources.

Since plants use carbon dioxide and sunlight for their carbon and energy sources, genetically modifying a plant to produce PHA would appear to be a more economical alternative to manufacturing PHA in bacteria. To this end, researchers engineered the enzymes controlling PHB synthesis into *Agrobacterium* plasmid, which was used to genetically transform *Arabidopsis thaliana*, a small flowering plant that is a member of the mustard (*Brassicaceae*) family, to produce polymers in plant tissues. Up to 14% PHB in dry weight of leaves has been achieved. The production of PHAs in plants has yet to be realized, as technical questions remain concerning the separation of polymers from plant tissue and the overall energy requirements of processing facilities.

Further Reading

Herbaceous Crops

Stout, B. A. and B. Cheze, eds. 1999. *CIGR Handbook of Agricultural Engineering, Vol. III: Plant Production Engineering*. St. Joseph, MI: American Society of Agricultural Engineers.

Wright, L. L. 1994. "Production technology status of crops." *Biomass and Bioenergy* 6:190–209.

Woody Crops

Nyland, R. D. 2002. *Silviculture: Concepts and Applications*, Second Edition. McGraw-Hill.

Stokes, B. J. and T. P. McDonald, eds. 1994. Proceedings of the International Energy Agency, Task IX, Activity 1 Symposium "Mechanization in Short Rotation, Intensive Culture Forestry" March 1–3. Mobile, AL: US Department of Agriculture, Forest Service, Southern Forest Experiment Station (available at www.woodycrops.org).

Wright, L. L. 1994. "Production technology status of crops." *Biomass and Bioenergy*, 6.

Storage

Stout, B.A. and B. Cheze, eds. 1999. *CIGR Handbook of Agricultural Engineering, Vol. III: Plant Production Engineering*. St. Joseph, MI: American Society of Agricultural Engineers.

Transgenic Crops

Galun, E. and A. Breiman. 1997. *Transgenic Plants*. Imperial College Press.

Lal, R. and S. Lal. 1993. *Genetic Engineering of Plants for Crop Improvement*. Boca Raton: CRC Press.

Somerville, C. R. 1994. "Production of Industrial Materials in Transgenic Plants." In *The Production and Uses of Genetically Transformed Plants*. M. W. Bevan, R. D. Harrison and C. J. Leaver, eds. London : Chapman and Hall.

Problems

4.1 Sketch the soil classification triangle. Note the most desirable soil classifications for growth of herbaceous and woody crops.

4.2 Production of both herbaceous and woody crops involves five steps. List these steps.

4.3 Characterize each of the following tillage methods. What are the advantages of each?

  (a) Conventional tillage

  (b) Conservation tillage

4.4 Forage crops are classified according to the method by which they are prepared for storage. For each of the following forage crops, indicate how they are harvested and stored:

    (a) Hay

    (b) Silage

4.5 Freshly harvested switchgrass is found to contain 30% moisture. Estimate the amount of energy required, in practice, to reduce moisture content to 10%. What fraction of the higher heating value of switchgrass does this energy represent?

4.6 List the major pieces of equipment employed in harvesting woody crops and indicate their functions.

4.7 Provide a definition for the term "transgenic crop."

4.8 List the six steps in creating a new transgenic crop, and provide a brief description of each step.

4.9 List three examples of techniques for achieving transformation of a host plant, and provide a brief description of each technique.

4.10 What is the purpose of including a selectable marker gene in a constructed transgene?

CHAPTER 5

Products from Biorenewable Resources

5.1 Introduction

Biorenewable resources can be transformed into a variety of products, including energy, transportation fuels, chemicals, and natural fibers. This chapter is an introduction to the various products that can be produced from biorenewable resources, and it gives some idea of their use in commercial markets. In subsequent chapters, the processes by which these transformations occur are detailed.

Bioenergy is defined as the conversion of biorenewable resources into heat or electric power. It is usually intended to designate energy for stationary applications as opposed to energy for transportation. This distinction is important as bioenergy applications typically employ solid or gaseous fuels, whereas transportation applications employ liquid fuels, although there are exceptions to these generalizations.

Transportation fuels are chemicals with sufficient energy densities (enthalpies of reaction per unit volume) and combustion characteristics to make them suitable for transportation applications. Generally, transportation fuels are confined to liquids such as alkanes, alcohols, and esters that are readily vaporized and burned within heat engines. Hydrogen may eventually be classified as a practical transportation fuel if gas storage technologies can substantially improve over present storage methods based on metal hydrides and high-pressure tanks. Wood and other solid biomass, despite respectable energy densities, do not have handling and combustion characteristics suitable for transportation fuels. The literature often fails to distinguish between *fuels*, which are any substances that can be burned for the production of heat or power, and *transportation fuels*, which is a distinct subset of fuels. The distinction is important because of the more exacting specifications required of transportation fuels.

Chemicals usually are taken to be commodity chemicals, such as acetic acid and propylene glycol, although it can include fine chemicals such as proteins and pharmaceutical chemicals. Commodity chemicals find applications in the manufacture of a wide variety of products including absorbents, adhesives, solvents, lubricants, inks, pesticides, coatings, films, and polymers. A chemical such as ethanol can be classified as either chemical or transportation fuel, depending on the ultimate application.

Natural fibers are the elongated cells in plant tissue that serve a structural function in plants. Very long, fine fibers such as cotton are used in the manufacture of cloth while wood fiber is the source of pulp in papermaking. Clearly, polymers synthesized from biorenewable resources can be spun into fibers, but these are synthetic fibers. In this book, fibers separated from biorenewable resources are classified as natural fibers whereas fibers spun from polymers that have been synthesized from biorenewable resources are classified as synthetic fibers.

The following sections describe the application and potential markets for products from biorenewable resources. The section on bioenergy describes the equipment used to convert thermal energy or gaseous fuel generated from biomass into process heat or electric power. The section on transportation fuels describes issues related to the satisfactory use of biofuels in transportation applications. The section on chemicals will introduce the reader to commodity chemicals in today's markets, and indicate where chemicals from biorenewable resources might find market entry. Finally, the section on fibers will explain how natural fibers, once separated from plant materials, can be manipulated into biobased products.

5.2 Bioenergy

Oxidation of carbon-carbon and carbon-hydrogen bonds in carbohydrates, lipids, and protein are exothermic reactions. Thus, if sufficiently dried, any biorenewable resource can be burned as a source of energy. In practice, sugars, starches, lipids and proteins are too valuable in the production of food, transportation fuels, and chemicals to be burned for heat and stationary power. Only lignocellulosic materials, such as wood and straw, and mixed waste streams, like manure or municipal solid wastes, are likely candidates for bioenergy since they are difficult to process into higher-value biobased products.

5.2.1 Process Heat

The simplest way to derive bioenergy from biorenewable resources is to burn it. The technical process of converting chemical energy into heat is described in Chapter 6. The utilization of this released energy for process heat and power is described in the following paragraphs.

Process heat is thermal energy used to raise the temperature of a process stream for the purpose of drying materials or promoting chemical reactions. Process heat may be required at temperatures as low as 30°C, for example, in heating an anaerobic digester, or as high as 2500°C, for example, in calcining limestone for the production of concrete. Fired process equipment, also known as a *furnace*, is used to

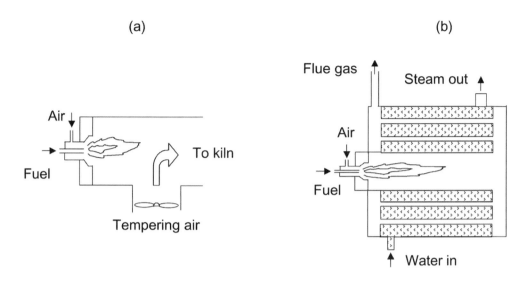

Fɪɢ. 5.1. Examples of process heaters (a) Direct-fired burner for a kiln, (b) Indirect-fired water-tube boiler.

burn fuel and deliver process heat at a desired temperature. As illustrated in Fig. 5.1, process heaters can be either direct fired or indirectly fired.

5.2.1.1 Direct-fired Furnaces

Direct-fired furnaces either burn fuel directly in the process stream, or the flue gas from combustion is brought into contact with the process stream. The most common examples of direct-fired furnaces are grain driers, wood kilns, cement kilns, and metal-forging ovens. As illustrated in Fig. 5.1(a), temperature control is achieved by adding air in excess of what is required to burn the fuel. Direct firing has the disadvantage of potentially introducing contaminants from the flue gas stream into the process stream. Although flue gas from the combustion of biomass is primarily a mixture of nitrogen, carbon dioxide, water vapor, and lesser amounts of oxygen, it also contains traces of alkali metals (sodium and potassium), acid gases (hydrogen chloride, sulfur dioxide, and nitrogen oxides), and tar, which is a mixture of oxygenated organic compounds. Modern standards of quality control preclude direct-fired furnaces fueled by biomass except in a few circumstances, such as wood drying or calcining of limestone. On the other hand, if the biomass is first converted to flammable gases, by either thermal gasification or anaerobic digestion, followed by a gas-cleaning step, the resulting gases may be suitable for use in direct-fired furnaces. However, gas cleaning results in additional capital cost. This cleaning step may be justified for advanced power generation, as subsequently described, but may not be economical for process heat applications.

5.2.1.2 Indirect-fired Furnaces

Indirect-fired furnaces do not allow the products of combustion to come in direct contact with the process stream. When very high-temperature process heat is required, the combustion stream is separated from the process stream by a thermally conducting wall. Such *process heaters* are commonly used for cracking hydrocarbons or as reboilers in distillation of heavy petroleum liquids.

Another approach to indirect-fired furnaces is to employ a heat-transfer medium to move thermal energy from the combustion stream to the process stream. *Air-to-air heat exchangers* have been successfully used in conjunction with biomass combustors to transfer heat from hot, dirty flue gas to a clean air stream. The relatively low convection coefficients associated with gas, however, require large heat exchangers. *Thermal fluid heaters* are employed in applications where high-pressure steam is to be avoided as a heat transfer medium. Suitable thermal fluids include water for relatively low temperature applications (100–250°C) and low vapor pressure organic liquids for moderate temperature applications (250–400°C). *Boilers* use steam as a clean, non-corrosive, high-energy content heat transfer medium capable of operating at moderately high temperatures (200–540°C). The steam generated is suitable for either direct-contact or indirect-contact heat exchange with process streams.

A *fire-tube boiler*, illustrated in Fig. 5.1(b), consists of a large, cylindrical shell containing water to be converted to steam. Through the center of this vessel passes a large diameter "fire tube" where combustion of the fuel occurs. The flue gas produced exits through a bundle of smaller tubes surrounding the central fire tube. These boilers are most suited for gaseous or volatile liquid fuels. *Water-tube boilers* consist of a large chamber, or firebox, that is either penetrated or surrounded by small-diameter tubes containing water. Fuel is burned in the firebox at high temperatures with heat transmitted to the steam tubes by radiation from the flame and by convection from the hot flue gases. Water-tube boilers are the most appropriate technology for direct combustion of biomass fuels. Raising steam from biomass fuels is examined in Chapter 6.

5.2.2 Stationary Power

A variety of cycles have been devised to convert heat into work. Power cycles of relevance to bioenergy include the Stirling, Otto, Diesel, Rankine, and Brayton cycles. From a theoretical perspective, the thermodynamic efficiencies of all five of these cycles are limited by the Carnot efficiency. This equation states that the maximum efficiency, η_{max}, of a heat engine is a function of the temperatures at which the engine receives heat from a high temperature, T_H, and rejects heat to a low temperature, T_L:

$$\eta_{max} = 1 - \frac{T_L}{T_H} \tag{5.1}$$

Accordingly, heat engines must operate at temperatures as high as can be tolerated by the materials of construction in order to achieve the maximum conversion of chemical energy into work. This limit, along with various irreversibilities such as friction, turbulence, and heat transfer, restricts thermodynamic efficiencies of practical heat engines to the range of 35–50%.

5.2.2.1 Stirling Cycle

The Stirling cycle is an example of an external combustion engine; that is, the products of combustion do not come into contact with the fluid that undergoes the thermodynamic processes of the cycle. In this respect, it resembles the Rankine steam cycle although its thermodynamic efficiency is theoretically higher than that of the Rankine cycle operating on the same fuel. In practice, the efficiencies of Stirling engines are relatively modest, as is their output power, typically no more than a few kilowatts. They continue to attract attention mainly because of their low maintenance requirements, high tolerance to contaminants, and relatively low pollution emissions. However, high costs have prevented significant market entry.

5.2.2.2 Otto and Diesel Cycles

Internal combustion engines are the practical manifestations of the Otto and Diesel cycles. They are robust in operation, efficient at small scales, and are more tolerant of contaminants than are gas turbines, as illustrated in Table 5.1. Nevertheless, they are generally confined to niche markets for stationary power applications. Probably the most important reason for this situation is a limit on their size, which is on the order of 5 MW. Economies of scale, which indicate that the cost of a product goes down as the production facility gets larger, have led the electric utility industry to build power plants that typically produce hundreds, if not thousands, of megawatts of electricity. Internal combustion engines cannot compete in this arena. Even at the

Table 5.1. Tolerance levels of contaminants in energy conversion equipment

Equipment	Particulate Loading (mg/Nm^3)	Alkali Loading (mg/Nm^3)	Tar Loading (g/Nm^3)
IC Engine	< 50	—	< 0.6
Turbocharged Engine	—	—	< 0.1
Gas Turbines:			
Aero-Derivative	< 50	< 0.24	< 0.008
Radial	< 40	—	—
Axial	< 13	< 0.24	—
Synthesis Gas	< 0.020	—	< 0.1

Source: S. P. Babu, 1995, "Thermal Gasification of Biomass Technology Developments: End of Task Report for 1992 to 1994," *Biomass and Bioenergy* 9:271-285.

Note: Nm^3 = normal cubic meters (evaluated at 1 atm, 298°K).

scale of hundreds of kilowatts, internal combustion engines may not be attractive for future bioenergy applications because of their relatively high pollution emissions and limited opportunities to enhance their efficiency. Application of internal combustion engines for vehicular propulsion is discussed in the section on transportation fuels.

5.2.2.3 Rankine Cycle

The Rankine cycle is another example of an external combustion cycle. As illustrated in Fig. 5.2, fuel is burned in a furnace where the released heat is transferred to pressurized water contained within steel tubes. Steam generated in this process is expanded in a steam turbine, which drives an electric generator to produce electric power. The steam is condensed in a water-cooled condenser, which controls the temperature exiting the turbine and thus sets T_L in Eq. 5.1.

The Rankine steam power cycle has been the foundation of stationary power generation for over a century. Although Brayton cycles employing gas turbines and electrochemical cycles based on fuel cells will constitute much of the growth in power generation in the future, steam power plants will continue to supply the majority of electric power for decades to come and will find new applications in combination with advanced generation technologies. The reason for the Rankine cycle's preeminence has been its ability to directly fire coal and other inexpensive solid fuels. Constructed at scales of several hundred megawatts, the modern steam

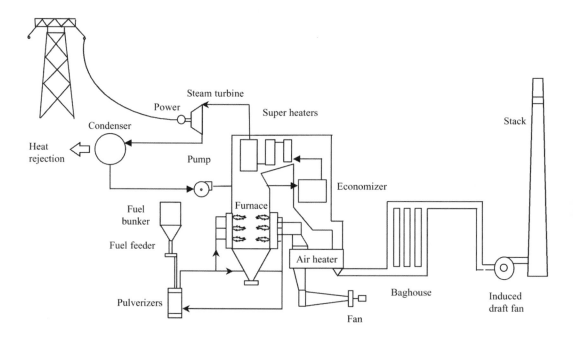

FIG. 5.2. Rankine steam power plant.

power plant can convert as much as 45% of chemical energy in fuel to electricity at a cost of $0.02–$0.05/kilowatt.

Utility-scale steam power plants are not expected to dominate future growth in electric power infrastructure in the United States. These giant plants take several years to plan and construct, which decreases their financial attractiveness in increasingly deregulated power markets. Coal and other fossil fuels burned in these plants are major sources of air pollution. They generate sulfur and nitrogen oxides, which are precursors to acid rain (the latter is also an important factor in smog formation); fine particulate matter, which is implicated in respiratory disease in urban areas; and heavy metals, the most prominent being mercury, which accumulates in the biosphere to toxic levels. Substitution of biorenewable resources, such as wood and agricultural residues, for coal in existing power plants could substantially reduce pollution emissions, although these plants are so large that the locally available biomass resources could supply only a small fraction of the total energy requirement. Small-scale steam power plants sized to use local biomass resources have low thermodynamic efficiencies, typically less than 25%, making them wasteful of energy resources.

5.2.2.4 Brayton Cycle

The Brayton cycle produces electric power by expanding hot gas through a turbine. These gas turbines operate at temperatures approaching 1300°C compared to inlet temperatures of less than 650°C for steam turbines used in Rankine cycles. Although this difference in inlet temperature (T_H in Eq. 5.1) would suggest that Brayton cycles have much higher thermodynamic efficiencies than Rankine cycles, the Brayton cycle also has much higher exhaust temperature (T_L in Eq 5.1) than does the Rankine cycle. Gas turbine exhaust temperatures are in the range of 400–600°C, whereas steam turbine exhaust temperatures are around 20°C. Furthermore, Brayton cycles, which contain both a gas compressor and gas turbine, have more sources of mechanical irreversibilities. This further degrades thermodynamic efficiencies, which may be only marginally higher than the best Rankine steam cycles. However, improvements in gas turbine technology that allow operation at higher temperatures and pressures are expected to increase Rankine cycle efficiency for large power plants to greater than 50%, although 30% is more realistic for gas turbines sized appropriately for biomass power plants.

The two general classes of gas turbines for power generation are heavy-duty industrial turbines and lightweight aeroderivative gas turbines. The *aeroderivatives* are gas turbines originally developed for commercial aviation but adapted for stationary electric power generation. They are attractive for bioenergy applications because of their high efficiency and low unit capital costs at the modest scales required for biomass fuels.

Gas turbines are well suited to gaseous and liquid fuels that are relatively free of contaminants that rapidly erode or corrode turbine blades. In this respect, gas turbine engines are not suitable for directly firing most biomass fuels. Solid biomass releases significant quantities of alkali metals, chlorine, mineral matter, and lesser amounts of sulfur upon burning. These would be entrained in the gas flow entering the expansion turbine where they would quickly contribute to blade failure. Cleaning large quantities of hot flue gas is not generally considered an economical proposition. Even the gas released from anaerobic digestion contains too much hydrogen sulfide to be directly burned in a gas turbine without first chemically scrubbing the gas to remove this corrosive agent.

Nevertheless, gas turbine engines are considered one of the most promising technologies for bioenergy because of the relative ease of plant construction, cost-effectiveness in a wide range of sizes (from tens of kilowatts to hundreds of megawatts), and the potential for very high thermodynamic efficiencies when employed in advanced cycles. The key to their success in bioenergy applications is converting the biomass to clean-burning gas or liquid before burning it in the gas turbine combustor. Chapter 6 explores technologies suitable for this purpose.

5.2.2.5 Combined Cycles

In an effort to enhance energy conversion efficiency, *combined cycle power systems* have been developed, which recognize that waste heat from one power cycle can be used to drive a second power cycle. Combined cycles would be unnecessary if a single heat engine could be built to operate between the temperature extremes of burning fuel (T_H) and the ambient environment (T_L). However, temperature and pressure limitations on materials of construction have prevented this realization. Combined cycles employ a topping cycle designed to operate between T_H and some intermediate temperature, T_I, and a bottoming cycle designed to operate between T_I and T_L. The overall efficiency of a combined cycle power system is:

$$\eta_C = \eta_T + \eta_B - \eta_T \eta_B \tag{5.2}$$

where η_C, η_T, and η_B are the efficiencies of the combined cycle, topping cycle, and bottoming cycle, respectively. Most commonly, combined cycle power plants employ a gas turbine engine for the topping cycle and a steam turbine plant for the bottoming cycle, achieving overall efficiencies of 60% or higher.

Clean-burning fuel from biomass for use in a combined cycle can be obtained by thermal gasification or anaerobic digestion to produce gas, or by fast pyrolysis to produce liquid. *Integrated gasification/combined cycle (IGCC) power* is illustrated in Fig. 5.3. Compressed air enters an oxygen plant, which separates oxygen from the air. The oxygen is used to gasify biomass in a pressurized gasifier to produce medium heating-value producer gas. The producer gas passes through cyclones and a gas clean-up system to remove particulate matter, tar, and other contaminants that may adversely affect gas turbine performance (alkali and chloride being the most prominent among these). These clean-up operations are best performed at high temperature and pressure to achieve high cycle efficiency. The clean gas is

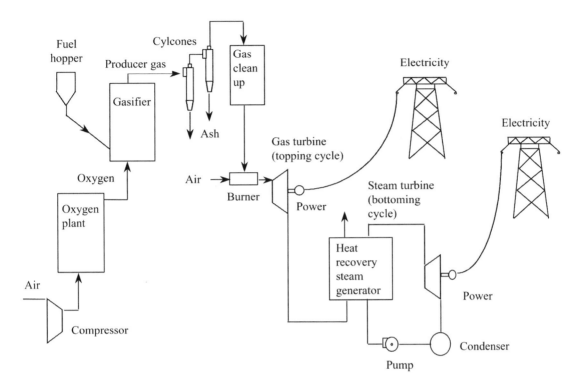

FIG. 5.3. Integrated gasification/combined cycle power plant (IGCC) based on a gas turbine topping cycle.

then burned in air and expanded through a gas turbine operating as a "topping" cycle. The gas exits the turbine at temperatures ranging between 400°C and 600°C. A heat recovery steam generator produces steam for a "bottoming" cycle that employs a steam turbine. Electric power is produced at two locations in this plant, yielding thermodynamic efficiencies exceeding 47%.

Integrated gasifier/combined cycle systems based on gas turbines are attractive for several reasons including their relative commercial readiness and the expectation that they can generate electricity at the lowest cost of all possible biomass power options.

An alternative to IGCC is to generate steam for injection into the gas turbine combustor, which increases mass flow and power output from the turbine. This variation, called a *steam-injected gas-turbine (STIG) cycle*, is less capital intensive than IGCC, since it does not employ a steam turbine. The STIG cycle is commercially developed for natural gas; lower flammability limits for producer gas make steam injection more problematic for biomass-derived producer gas. The *intercooled steam-injected gas turbine (ISTIG)* is an advanced version of the STIG.

This cycle further improves thermodynamic efficiency by cooling gas flow between several stages of compression, a process known as intercooling.

5.2.2.6 Fuel Cells

Among the most exciting new energy technologies are *fuel cells*, which directly convert chemical energy into work, thus bypassing the restriction on efficiency imposed by the Carnot relationship. This does not imply that fuel cells can convert 100% of the chemical enthalpy of a fuel into work, as the process still must conform to the laws of thermodynamics. Thermodynamic efficiency, as defined in Chapter 2, is the ratio of useful energy out to the total energy into the process. For a fuel cell, the useful energy out is the electrical work generated. From chemical thermodynamics it is known that the maximum theoretical work for a chemical reaction is equal to the change in Gibbs function, ΔG, for the reaction. The total chemical energy entering the fuel cell is the enthalpy of reaction, ΔH. Therefore, the efficiency of a fuel cell operating at 1 atmosphere of pressure is:

$$\eta_{max} = \frac{E_{out}}{E_{in}} = \frac{\Delta G^o}{\Delta H} = \frac{\Delta H - T\Delta S^o}{\Delta H} = 1 - T\frac{\Delta S^o}{\Delta H} \tag{5.3}$$

where the superscript indicates calculation at 1 atmosphere of pressure. All practical fuel cells to date are based on the oxidation of hydrogen:

$$H_2 + \tfrac{1}{2}O_2 \rightarrow H_2O \tag{5.4}$$

This reaction has an enthalpy of $-285,830$ kJ/kmol and entropy of reaction of -163 kJ/kmol-K at 25°C. Thus, a fuel cell operating at 25°C (298 K) has a theoretical maximum efficiency of:

$$\eta_{max} = \left[1 - 298 \text{ K} \frac{(-163 \text{ kJ/kmol-K})}{(-285,830 \text{ kJ/kmol})} \right] \times 100 = 83\% \tag{5.5}$$

Irreversibilities acting on the fuel cell reduce this efficiency to 35–60%, depending on the fuel cell design. Thus, fuel cells can produce significantly more work from a given amount of fuel than can heat engines. However, carbonaceous fuels must first be reformed to hydrogen before they are suitable for use in fuel cells. The energy losses associated with fuel reforming must be included when determining the overall fuel-to-electricity conversion efficiency of a fuel cell.

The operation of a fuel cell is illustrated schematically in Fig. 5.4. The device consists of two gas-permeable electrodes separated by an electrolyte, which is a transport medium for electrically charged ions. Hydrogen gas, the ultimate fuel in all current designs of fuel cells, enters the fuel cell through the anode while oxygen is admitted through the cathode.

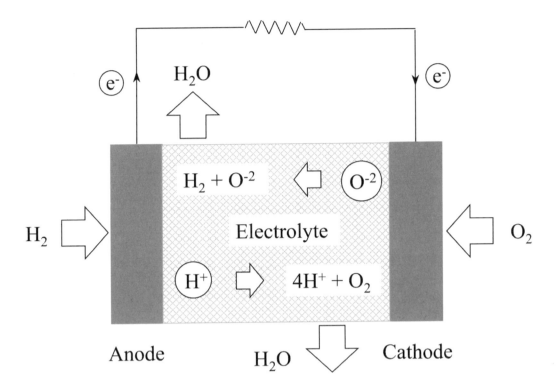

FIG. 5.4.   Schematic of operation of fuel cell.

Depending on the fuel cell design, either positively charged hydrogen ions form at the anode or negatively charged ions containing oxygen form at the cathode. In either case, the resulting ions migrate through the electrolyte to the opposite electrode from which they are formed. Hydrogen ions migrate to the cathode where they react with oxygen to form water. Oxygen-bearing ions migrate to the anode where they react with hydrogen to form water. Both ionic processes release chemical energy in the form of electrons at the anode, which flow to the cathode through an external electric circuit. The flow of electrons from anode to cathode represents the direct generation of electric power from flameless oxidation of fuel. The inherently high thermodynamic efficiency of fuel cells make them attractive for biomass power where fuel costs are relatively expensive compared to many fossil fuels.

Several types of fuel cells have been developed and they are classified according to the kind of electrolyte employed: phosphoric acid, polymeric, molten carbonate, and solid oxide. Despite differences in materials and operating conditions, all are based on the electrochemical reaction of hydrogen and oxygen. For integrated gasification/fuel cell power systems, the molten carbonate and solid oxide systems are of particular interest because of their high temperature operation, which allows heat recovery and use in advanced power cycles.

The first generation of fuels cells employs *phosphoric acid* as the electrolyte. Operated at 200°C, it attains efficiencies of 35–45%. Although well developed, commercial application has been limited by its use of expensive platinum catalysts, which are deactivated by CO.

Proton exchange membrane fuel cells were originally developed for space applications but are also considered attractive for automotive propulsion. The electrolyte is a solid polymeric material operated at less than 100°C; both factors contribute to the ease of construction and operation of this fuel cell. Efficiencies range from 35 to 60%. However, this fuel cell is intolerant of CO, which can be difficult to exclude from on-board reformed fuels. It would also require use of "cold" gas, which reduces the efficiency of a gasification-based power system.

The *molten carbonate fuel cell* operates at about 650°C. Although hydrogen is the ultimate energy carrier, this fuel cell can be operated on a variety of hydrogen-rich fuels, including methane, kerosene, diesel fuel, ethanol, and producer gas. Within the fuel cell, a reformer converts these fuels into mixtures of hydrogen, carbon monoxide, carbon dioxide, and water along with varying amounts of unreformed fuel.

Overall efficiency of converting methane into electricity is 45–55%. Molten carbonate fuel cells have completed extensive demonstration trials but they have unresolved materials problems related to the use of corrosive alkali carbonate as electrolyte.

The *solid oxide system* is also a high temperature fuel cell, operating at 650–1000°C. The electrolyte is a solid, non-porous ceramic material, usually based upon yttrium and zirconium. The solid electrolyte provides for a simpler, less expensive design and longer expected life than other fuel cell systems. Efficiencies range from 50 to 60%. The higher operating temperature compared to the molten carbonate system also enhances its attractiveness for heat recovery and use in advanced power systems. Most solid oxide fuel cells operate with a steam reformer in which water vapor and carbon monoxide react according to the water-gas shift reaction to produce hydrogen, thus making them compatible with carbon monoxide-rich producer gas as fuel. Solid oxide systems may solve some of the corrosion problems associated with molten carbonate fuel cells.

The gas mixture produced by a biomass gasifier contains dust and tar that must be removed or greatly reduced for most applications, including power generation in fuel cells. Removal of tar would ideally be performed at elevated temperatures. If the gas is to be used in fuel cells, further cleaning is required to remove ammonia (NH_3), hydrogen chloride (HCl), and hydrogen sulfide (H_2S). Table 5.2 details contaminant removal requirements for various kinds of fuel cells. To obtain high-energy efficiency, trace contaminant removal must be performed at elevated temperatures for fuel cells that operate at relatively high temperatures. Low

Table 5.2. Contaminants for various fuel cell systems

Fuel Cell Type	Classification	Contaminant Tolerance Level			
		H_2S	HCl	NH_3	CO
Solid polymer	Low temperature	—	—	—	10 ppm
Phosphoric acid	Low temperature	50 ppm	—	—	5%
Molten carbonate	High temperature	0.5 ppm	10 ppm	1%	—
Solid oxide	High temperature	0.1 ppm	1.0 ppm	0.5%	—

Source: *Fuel Cell Handbook*, Fifth Edition, Oct. 2000 (CD-ROM) (Morgantown, WV/ Pittsburgh, PA: US Department of Energy National Energy Technology Laboratory).

temperature fuel cells cannot tolerate CO, which can be removed by the water-gas shift reaction. The catalysts that facilitate the shift reaction, however, are poisoned by trace contaminants, which must be removed prior to the shift reactors. One method for removing H_2S and HCl is the use of a fixed bed of calcined dolomite or limestone and zinc titanate at temperatures around 630°C. This is followed by steam reforming at high temperature (750 to 850°C) to destroy tar and ammonia.

Figure 5.5 illustrates an IGCC power plant based on a molten carbonate fuel cell. Biomass is gasified in oxygen to yield producer gas. Gasification occurs at elevated pressure to improve the yield of methane, which is important for proper thermal balance of this fuel cell. Hot-gas clean up to remove particulate matter, tar, and other contaminants is followed by expansion through a gas turbine as part of a topping power cycle. The pressure and temperature of the producer gas is sufficiently reduced after this to admit it into the fuel cell. High temperature exhaust gas exiting the cathode of the fuel cell enters a heat recovery steam generator, which is part of a bottoming cycle in the integrated plant. Thus, electricity is generated at three locations in the plant for an overall thermodynamic efficiency reaching 60% or more.

5.3 Transportation Fuels

Almost 25% of energy consumption in the United States is consumed by transportation needs. Approximately half of this amount comes from imported petroleum. Thus, development of transportation fuels from biorenewable resources is a priority if decreased dependence on foreign sources of energy is to be achieved. In principle, the United States has sufficient biorenewable resources to satisfy all of its transportation needs.

Traditional transportation fuels are classified as gasoline, diesel fuel, or jet fuel. Gasoline is intended for spark-ignition (Otto cycle) engines; thus, it is relatively volatile but resistant to autoignition during compression. Diesel fuel is intended for use in compression-ignition (Diesel cycle) engines; thus, it is less volatile compared to gasoline and more susceptible to autoignition during compression. Jet fuel

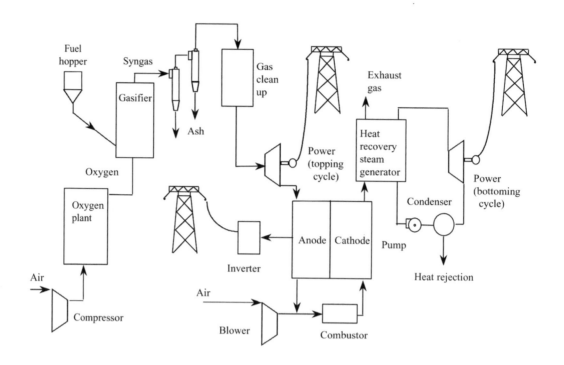

FIG. 5.5. Combined-cycle power plant based on high temperature fuel cells.

is designed for use in gas turbine (Brayton cycle) engines, which are not limited by autoignition characteristics but otherwise have very strict fuel specifications for reasons of safety and engine durability.

Gasoline is a mixture of hundreds of different hydrocarbons that contain from 4 to 12 carbon atoms with boiling points in the range of 25–225°C and are obtained from a large number of refinery process streams. Most of the mixture consists of alkanes, with butanes and pentanes added to meet vapor pressure specifications. A few percent of aromatic compounds are added to increase *octane number*, the figure of merit used to indicate the tendency of a fuel to undergo premature detonation within the combustion cylinder of an internal combustion engine. The higher the octane number, the less likely a fuel will detonate until exposed to an ignition source (electrical spark). Premature denotation is responsible for the phenomenon known as engine knock, which reduces fuel economy and can damage an engine. Various systems of octane rating have been developed, including research octane and motor octane numbers. Federal regulation in the United States requires gasoline sold commercially to be rated using an average of the research and motor octane numbers. Gasoline rated as "regular" has a commercial octane number of about 87 while premium grade is 93.

Diesel fuel, like gasoline, is also a mixture of light distillate hydrocarbons but with lower volatility and higher viscosity. Because diesel fuel is intended to be ignited by compression rather than by a spark, its autoignition temperature is lower than for gasoline. The combustion behavior of diesel fuels is conveniently rated according to *cetane number*, an indication of how long it takes a fuel to ignite (ignition delay) after it has been injected under pressure into a diesel engine. A high cetane number indicates short ignition delay; for example, No. 2 diesel fuel has a cetane number of 37–56 while gasoline has a cetane number less than 15.

Jet fuel is designated as either Jet A fuel or Jet B fuel. Jet A fuel is a kerosene type of fuel with relatively high flash point whereas Jet B fuel is a wide boiling range fuel, which more readily evaporates.

The primary candidates for biobased transportation fuels are ethanol, methanol, and biodiesel. Ethanol and methanol, by virtue of their high octane numbers, are suitable for use in spark-ignition engines, while the high cetane numbers of biodiesel, which are methyl or ethyl esters formulated from vegetable or animal fats, make them suitable for use in compression-ignition engines. Properties of fossil fuel-derived gasoline and diesel fuel are compared to biorenewable resource-derived methanol, ethanol, and methyl ester in Table 5.3.

5.3.1 Ethanol

Ethanol can be produced by the fermentation of sugar or starch crops. A fuel ethanol market has been developed in Brazil based on sugar cane while the United

Table 5.3. Comparison of transportation fuels

Fuel type	Fossil Fuel-derived		Biorenewable Resource-derived		
	Gasoline	No. 2 Diesel Fuel	Methanol	Ethanol	Methyl ester (from soybean oil)
Specific gravity at 16°C	0.72–0.78	0.85	0.796	0.794	0.886
Kinematic viscosity at 20°C (m/s)	0.8×10^{-6}	2.5×10^{-6}	0.75×10^{-6}	151×10^{-6}	3.9×10^{-6}
Boiling point range (°C)	30–225	210–235	65	78	339
Flash point (°C)	−43	52	11	13	188
Autoignition temperature (°C)	370	254	464	423	—
Octane no. (research)	91–100	—	109	109	—
Octane no. (motor)	82–92	—	89	90	—
Cetane no.	<15	37–56	<15	<15	55
Heat of vaporization (kJ/kg)	380	375	1185	920	—
Lower heating value (MJ/kg)	43.5	45	20.1	27	37

Source: G. L. Borman and K. W. Ragland, 1998, *Combustion Engineering* (McGraw-Hill).

Note: MJ = megajoules.

States has relied on cornstarch for commercial production of fuel ethanol. Technologies are also being developed to convert lignocellulose into sugars or syngas, a mixture of carbon monoxide and hydrogen, either of which can be fermented into ethanol.

On a volumetric basis, ethanol has only 66% of the heating value of gasoline. Thus, the range of a vehicle operating on pure ethanol is theoretically reduced by a corresponding amount. Accordingly, meaningful comparisons between the cost of gasoline and the cost of ethanol should be made on the basis of energy delivered ($/GJ) instead of fuel volume ($/liter). However, fuel economy depends on many complex interactions between a fuel and the combustion environment within an engine that some argue improves the relative performance of ethanol. For example, as Table 5.3 illustrates, ethanol has a much tighter boiling point range and higher latent heat of vaporization compared to gasoline. This promotes evaporation of fuel droplets during the compression stroke of an engine, cooling the engine, and allowing higher densities of fuel and air to be burned in the engine. Furthermore, ethanol, with molecular formula of C_2H_5OH, has a much lower stoichiometric air-fuel ratio than gasoline, nominally taken to have the molecular formula of octane (C_8H_{18}), and so a higher energy density can be achieved in the engine. These factors result in higher inducted fuel energy densities, which can produce both higher power outputs and improved engine efficiency compared to gasoline-fueled vehicles. Finally, the higher octane number for ethanol compared to gasoline (109 vs. 91–101) allows engines to be designed to run at higher compression ratios, which improves both power and fuel economy. Estimates for efficiency improvements in engines optimized for ethanol instead of gasoline range from 15 to 30%, resulting in a driving range approaching 80% of that of gasoline.

Internal combustion engines can be fueled on pure ethanol (known as neat alcohol or E100) or on blends of ethanol and gasoline. In Brazil they employ 190 proof ethanol (95 vol % alcohol and 5% water), which eliminates the energy-consuming step of producing anhydrous ethanol. In the United States two ethanol-gasoline blends are common: E85 contains 85% ethanol and 15% gasoline, and E10, also referred to as gasohol, contains only 10% ethanol with the balance being gasoline. The advantage of E10 is that it can be used in vehicles with engines designed for gasoline; however, its use is accompanied by a loss in fuel economy (as measured in km/liter or miles/gal) compared to gasoline, amounting to 2–5%.

A significant problem with ethanol-gasoline blends is water-induced phase separation. Water contaminating a storage tank or pipeline is readily absorbed by ethanol, resulting in a lower water-rich layer and an upper hydrocarbon-rich layer that interferes with proper engine operation. Water contamination is a problem that has not been fully addressed by the refining, blending, and distribution industries; thus transportation of ethanol-gasoline blends in pipelines is not permitted in the United States and long-term storage is to be avoided.

5.3.2 Methanol

Methanol is among the top ten commodity chemicals produced in the United States. Much of it is used to produce 2-methoxy-2-methylpropane (commonly known as methyl-tertiary-butyl ether or MTBE), an oxygenate and octane enhancer in unleaded gasoline. Although destructive distillation of wood was the original source for methanol, today it is produced almost exclusively from syngas derived from natural gas.

The fuel properties of methanol are similar to those of ethanol: narrow boiling point range, high heat of vaporization, and high octane number. It has only 49% of the volumetric heating value of gasoline. As a transportation fuel, it has many of the same advantages and disadvantages as ethanol. Nevertheless, the prospects for methanol as a biobased product are not promising due to economic and environmental factors. First, the route from bioresource to methanol is the same as for fossil fuel to methanol: steam reforming or gasification of feedstock to *syngas*, a mixture of carbon monoxide and hydrogen, followed by catalytic conversion of the syngas at high pressure to yield methanol. Since biorenewable resources are typically more expensive than fossil resources, it is unlikely that they will be substituted for fossil fuels to produce methanol. Furthermore, methanol is considerably more toxic than ethanol. Recent rulings by the U.S. Environmental Protection Agency (EPA) are likely to ban the closely related and similarly toxic MTBE as a fuel additive because of concerns about ground water contamination.

Methanol-gasoline blends have much lower tolerance to water than do ethanol-gasoline blends. The EPA has limited the blending of methanol without cosolvent to less than 0.3 vol % because of phase separation problems.

5.3.3 Biodiesel

Vegetable oils, which are triglycerides of fatty acids, have long been recognized as potential fuels in diesel engines. Compared to petroleum-based diesel fuels, vegetable oils have higher viscosity and lower volatility resulting in fouling of engine valves and less favorable combustion performance, especially in direct-injection engines. The solution to this problem is to convert the triglycerides into methyl esters or ethyl esters of the fatty acids, known as biodiesel, and the by-product 1,2,3-propanetriol (glycerol).

Table 5.3 illustrates that fuel properties of biodiesel are very similar to petroleum-based diesel. The specific gravity and viscosity of biodiesel are only slightly higher than for diesel while the cetane numbers and heating values are comparable. Significantly higher flash points for biodiesel represent greater safety in storage and transportation. Biodiesel can be used in unmodified diesel engines with no excess wear or operational problems. Tests in light and heavy trucks showed few differences other than a requirement for more frequent oil changes because of the build-up of ester fuel in engine crankcases.

5.4 Chemicals

Chemicals are the broadest class of biobased products. It would be impossible to describe all possible chemical products from biorenewable resources in this section. Instead, an effort is made to focus on potential commodity chemicals from biorenewable resources although fine chemicals and pharmaceuticals are also potential products. Table 5.4 lists the top sixty commodity chemicals in the United States in terms of their annual production. The table includes the feedstocks employed, and the energy consumed, in the production of the chemicals, as well as their selling prices in 1995. (Prices of commodity chemicals can fluctuate widely, even during the course of a year, so current market reports should be consulted if accurate prices are important.)

In principle, all organic compounds can be synthesized from biorenewable resources. In practice, current economics favors chemical synthesis from fossil resources, especially from petroleum and natural gas, except for a few oxygenated organic compounds. The following three sections summarize three categories of potential biobased chemicals: commercially important chemicals that are currently produced from fossil resources but could also be produced from biorenewable resources, those that are currently produced from biorenewable resources, and unexploited chemicals with future potential in the manufacture of biobased products.

5.4.1 Chemicals Produced from Fossil Resources

The foundation of the petrochemical industry consists of seven basic building blocks: syngas from methane (CH_4), ethylene (ethene – C_2H_4), propylene (propene – C_3H_6), butanes (C_4H_{10}), butylene (butene – C_4H_8), butadiene (C_4H_6), and BTX, which is a mixture of the closely related compounds benzene, toluene (methylbenzene), and xylene (dimethylbenzene). Methane is obtained from natural gas. Ethylene, propylene, butane, butylene, and butadiene are obtained from natural gas liquids, refinery off-gases, and petroleum. Benzene, toluene, and xylene, the basis for aromatic compounds, are obtained from petroleum and coal. From these seven building blocks all the organic compounds of the modern world can be obtained.

The following paragraphs describe several industrially important organic compounds, currently obtained from fossil resources, that might be obtained in the future from biorenewable resources. Some prospective methods for converting biorenewable resources into these chemicals are explored in Chapter 7.

Ethylene

Ethylene is the leading organic chemical manufactured in the United States and is the starting point for a wide variety of consumer and industrial products includ-

Table 5.4. Sixty major organic chemicals manufactured in the United States

No.	Product	Annual Production (10^9 kg)	Usual Organic Feedstock	Energy Consumed per Product Mass		Average Market Price ($/kg)
				Process (MJ/kg)	Feed (MJ/kg)	
1.	Ethylene	21.30	Natural gas liquids	23.3	58.1	0.55
2.	Propylene	11.65	Natural gas liquids	16.3	56.8	0.42
3.	MTBE	7.99	C_4H_8/methanol	13.0	50.5	0.29
4.	Ethylene dichloride	7.83	C_2H_4	4.0	42.2	0.37
5.	Benzene	7.24	BTX	2.3	81.2	0.33
6.	Urea	7.07	$CO_2(NH_3)$	2.8	30.6	0.20
7.	Vinyl chloride	6.79	C_2H_4	17.4	53.4	0.45
8.	Ethylbenzene	6.19	C_2H_4/benzene	2.3	66.5	0.56
9.	Styrene	5.16	Ethylbenzene	12.8	74.3	0.64
10.	Methanol, synthetic	5.12	$CO/(H_2)$	11.6	25.8	0.21
11.	Mixed xylenes	4.25	BTX	10.2	43.0	0.31
12.	Ethanol, fermentation	3.90	Corn	18.8	31.0	0.42
13.	Formaldehyde (37 wt %)	3.68	Methanol	0.0	43.9	0.24
14.	Terephthalic acid	3.61	p-Xylene	12.8	57.8	0.88
15.	Ethylene oxide	3.46	C_2H_4	0.4	68.1	0.99
16.	Toluene	3.05	BTX	10.2	43.0	0.31
17.	p-Xylene	2.88	BTX	22.3	59.1	0.46
18.	Cumene	2.55	C_3H_6/benzene	2.3	66.2	0.51
19.	Ethylene glycol	2.37	Ethylene oxide	13.9	49.6	0.37
20.	Acetic acid, synthetic	2.12	Methanol/(CO)	10.7	27.7	0.79
21.	Phenol, synthetic	1.89	Cumene	15.1	99.3	0.90
22.	Propylene oxide	1.81	Propylene	16.3	58.4	1.41
23.	1,3 Butadiene	1.67	Butanes/enes	51.1	58.4	0.49
24.	Carbon black	1.51	Residual oil	27.9	93.0	0.64
25.	Isobutylene	1.47	Butanes/enes	9.3	85.3	0.68
26.	Acrylonitrile	1.45	Propylene	6.0	78.8	1.17
27.	Vinyl acetate	1.31	C_2H_4/HOAc	15.6	59.2	0.97
28.	Acetone	1.25	Cumene	15.1	99.3	0.86
29.	Buyraldehyde	1.22	Propylene/(CO)	—	—	0.95
30.	Cyclohexane	0.969	Benzene	2.3	58.1	0.44
31.	Adipic acid	0.816	Cyclohexane	29.1	84.0	1.62

continues

127

Table 5.4. (Continued)

No.	Product	Annual Production (10^9 kg)	Usual Organic Feedstock	Energy Consumed per Product Mass Process (MJ/kg)	Energy Consumed per Product Mass Feed (MJ/kg)	Average Market Price ($/kg)
32.	Nitrobenzene	0.748	Benzene	—	—	0.73
33.	Bisphenol A	0.736	Acetone/phenol	—	—	2.07
34.	Caprolactam	0.714	Cyclohexanone	7.9	54.5	2.05
35.	Acrylic acid	0.698	Propylene	—	—	1.92
36.	n-Butanol	0.677	Propylene (CO)	—	—	1.10
37.	Isopropyl alcohol	0.646	Propylene	—	—	0.71
38.	Aniline	0.631	Nitrobenzene	—	—	1.08
39.	Methyl methacrylate	0.622	Acetone/(HCN)	—	—	1.32
40.	Cyclohexanone	0.501	Phenol	—	—	1.61
41.	Methyl chloride	0.483	Methane	—	—	0.85
42.	o-Xylene	0.465	BTX	—	—	0.42
43.	Propylene glycol	0.462	Propylene oxide	—	—	1.43
44.	Phthalic anhydride	0.451	Naphthalene	—	—	0.84
45.	Acetone cyanohydrin	0.410	Acetone/(HCN)	—	—	0.77
46.	Toluene diisocyanates	0.395	Toluene	—	—	2.09
47.	Dodecylbenzene	0.386	C_6H_6/dodecene	—	—	1.25
48.	Ethanol amines	0.369	Ethylene oxide	—	—	1.26
49.	Diethylene glycol	0.354	Ethylene oxide	—	—	0.71
50.	Carbon tetrachloride	0.344	Methane	—	—	0.79
51.	2-Ethyl-1-hexanol	0.337	C_3H_6/CO or RCHO	—	—	1.23
52.	Ethanol, synthetic	0.284	Ethylene	4.7	31.4	0.42
53.	Isoprene	0.281	HC/turpentines	—	—	0.68
54.	1,4-Butanediol	0.266	C_3H_6/CO or THF	—	—	1.24
55.	Methyl ethyl ketone	0.264	2-Butanol	—	—	1.01
56.	Ligninsulfonic acid salt	0.258	Sulfite liquor	—	—	0.36
57.	Chloroform	0.256	Methane	—	—	0.87
58.	Maleic anhydride	0.251	Benzene	—	—	0.93
59.	C_{12}-benzenesulfonate Na	0.217	C_6H_6/dodecene	—	—	1.34
60.	1-Butene	0.209	Ethylene	—	—	0.57

Source: D. L. Klass, 1998, *Biomass for Renewable Energy, Fuels, and Chemicals* (San Diego: Academic Press).

ing polymers (polyethylene, polystyrene, and polyester) and polyols (ethylene glycol). It is commercially produced from natural gas liquids.

Ethylene can also be produced by dehydration of ethanol in the presence of an acid catalyst. Thus, some have suggested the conversion of inexpensive lignocellulosic biomass into ethanol followed by the manufacture of biobased ethylene. Although Brazil built a commodity chemical industry on this process, the low cost of natural gas makes this an unlikely prospect in other markets for the foreseeable future.

1,3-Butadiene

Sixty percent of 1,3-butadiene (C_4H_6) production in the United States is used to produce elastomers such as styrene/butadiene, polybutadiene, neoprene, and nitrile rubber. The largest single market for butadiene is in the manufacture of tires. Smaller amounts go into other consumer and industrial products. Butadiene is manufactured in the petroleum industry by thermal cracking of natural gas liquids.

Pentane/Pentene

Pentane (C_5H_{12}) is used as a solvent, as a blending component of high-octane gasoline, and in processing isoprene. Pentene (C_5H_{10}) is used in the formulation of plasticizers and detergents and is a precursor to the production of thiols, amines, and ammonium salts. Pentenes have been used as monomers for the manufacture of resins and low molecular weight thermoplastic materials. Thermal cracking of naphtha and gas oil yields ethylene and propylene as the primary products and five carbon hydrocarbons as by-products, from which pentane and pentene are distilled.

Benzene

The single largest use of benzene (C_6H_6) in the United States is in the production of styrene, which finds application in the production of such polymers as polystyrene, acrylonitrile/butadiene/styrene (ABS), styrene/acrylonitrile resins, and styrene/butadiene latexes. Benzene is also used in the manufacture of cumene (isopropyl benzene), a chemical used in the production of 96% of domestic phenol; and cyclohexane, important in the manufacture of nylon. Small amounts find uses in the manufacture of detergents, insecticides, antioxidants, adhesives, and pharmaceuticals. Benzene is extracted from the BTX fraction obtained from catalytic reforming of naphtha.

Toluene

More than 90% of toluene (methylbenzene) is directed to the production of gasoline. Of the remaining amount, over half goes to the production of benzene, while smaller amounts are used as a solvent in the production of paints and coatings and other miscellaneous products. Toluene is extracted from the BTX fraction obtained from catalytic reforming of naphtha.

Xylenes

Mixed xylenes (dimethylbenzene) consist of ortho-, meta-, and para- isomers. Separation of these isomers provides starting material for phthalic anhydride, used in plasticizers; isophthalic acid, a component of polyester resins; and dimethyl terephthalate and terephthalic acid, intermediates in production of polyterephthalates. Mixed xylenes are also employed as solvents and in the formulation of adhesives and rubber. Xylenes are extracted from the BTX fraction obtained from catalytic reforming of naphtha.

Acetic Acid

Over sixty percent of acetic acid (ethanoic acid) produced is used in the manufacture of vinyl acetate, the basis of white glue, laminating wallboard, and latex paint. Polymers derived from vinyl acetate find applications in the manufacture of safety glass, film products, and hot-melt adhesives. About 15% is consumed in the production of acetic anhydride, most of which is converted to cellulose acetate, which is used to manufacture textile fibers, plastics, and films. About 10% of acetic acid demand supports production of ester solvents used in the manufacture of inks, paints, and coatings. Another 10% is used to produce terephthalic acid (1,4-benzene dicarboxylic acid) and dimethyl terephthalate (dimethyl 1,4-benzene dicarboxylate) for the manufacture of fibers, resins, paints, coatings, and plastics.

The production of acetic acid from fossil fuels begins with the reforming of natural gas or gasification of coal to syngas. As described in Chapter 7, syngas is catalytically converted to methanol followed by catalytic carbonylation to acetic acid, that is, synthesis of a carbonyl compound by the addition of carbon monoxide.

Acetic acid is a metabolite from the fermentation of a variety of sugars by several organisms, including *Acetobacter aceti, Clostridium thermoaceticum,* and *Pachysolen tannophilus.* However, acetic acid from fermentation is not cost-effective compared to production from fossil resources.

Formic Acid

Demand for formic acid (methanoic acid) is evenly distributed in the manufacture of rubber chemicals, aluminum and nickel catalysts, leather and tanning, textiles, and other miscellaneous industries. Partial oxidation of butane, obtained from natural gas liquids, yields a mixture of carboxylic acids from which formic acid is separated.

Formic acid and three other carboxylic acids (acetic, glyceric, and lactic) can be produced by mild alkaline oxidation of dilute aqueous solutions of xylose, representing a possible path for production of biobased products. However, the recovery and commercial exploitation of carboxylic acids are impeded by the fact that they are highly water-soluble.

Propionic Acid

Up to 30% of propionic acid (propanoic acid) is used to preserve grain. Another 25% is converted to sodium and calcium salts of the acid and are used as food and feed preservatives. Production of carboxylic acid herbicides consumes another 15%, with the balance used for cellulose plastics, pharmaceuticals, and solvents. Propionic acid is obtained by carbonylation of ethylene, which is derived from the pyrolysis of ethane, a constituent of natural gas.

Propionic acid is a metabolite from the fermentation of various sugars by various microorganisms including *Clostridium* species and *Propionibacterium shermanii.* However, the commercial production of propionic acid by fermentation is not economically viable at present.

Acrylic Acid

Acrylic acid (2-propenoic acid) is an unsaturated liquid acid that is widely used to produce acrylic esters, which are useful in the production of latexes, textiles, adhesives, sealants, and inks. Acrylic esters, which command 72% of production in acrylic acid, have increasing applications as cross-linking agents for producing hard, cured surfaces. Non-ester polymers of acrylic acid have applications as flocculants, dispersants, and thickeners. New types of polyacrylic acids have recently been developed as superabsorbants such as are used in disposable diapers. A two-stage oxidation process is used to produce acrylic acid from propylene.

Succinic Acid

Succinic acid (butanedioic acid) is a crystalline dicarboxylic acid found as a common metabolite in many plants, animals, and microorganisms. It is a specialty chemical with applications in food and pharmaceutical products, surfactants and detergents, biodegradable solvents and plastics, and ingredients to stimulate animal and plant growth. Succinic acid is produced from fossil resources by oxidation of butane to maleic anhydride (2,5-furandione), followed by reduction to succinic anhydride (dihydro-2,5-furandione) and ring-opening hydrolysis to give succinic acid.

However, to qualify as a biobased product it must be produced from biorenewable resources rather than from fossil fuels. Efforts are under way to develop a commercial process for fermenting it from cornstarch.

Butanol

Butanol is a four-carbon alcohol (C_4H_9OH) of either of two isomeric forms (1-butanol or 2-butanol) derived from normal butane. Butanol has been suggested as an oxygenated fuel for blending with gasoline, with distinct advantages over methanol and ethanol. The energy content of butanol is closer to that of gasoline. It has no compatibility or miscibility problems with gasoline. It is water tolerant and so can be transported in gasoline blends by pipeline without danger of phase separation due to moisture absorption. Butanol is also a feedstock for production of butyl butyrate, often touted as a green solvent.

Butanol is obtained from butane, which is extracted from natural gas liquids. However, various species of *Clostridium* can ferment butanol from sugars. Like many other fermentation processes, recovery of the metabolite is a key challenge to the economics of biobased butanol.

Phenol

About 35% of phenol production goes into phenolic resins, which are primarily employed as adhesives for plywood, laminates, and insulation. About 10% of these resins are used in molding compounds for heat-resistant materials. Another 35% of phenol production supports the manufacture of polycarbonate resins, a common material in the construction of automobiles, appliances, and milk bottles; and epoxy resins, a surface coating for electronic circuit boards and a component of advanced composites. Smaller amounts of phenol are employed to manufacture nylon carpet and aniline, a useful chemical in the manufacture of inks, dyes, herbicides, and other products. Phenol is used in the production of salicylic acid, the active component of aspirin and flavorings. Phenol from petrochemicals is derived by oxidizing cumene (isopropyl benzene), which is synthesized from benzene and propylene.

5.4.2 Chemicals Produced from Biorenewable Resources

Several oxygenated organic compounds are commercially produced from biorenewable resources. In large part these biobased products can compete in the market because synthesis routes from fossil resources to yield a comparable product have not emerged. The exceptions are lactic acid, for which the "natural" product is competitive with the "synthetic" one, and some cellulose derivatives, for which consumers value the intrinsic properties of the resulting biobased synthetic fibers.

Furfural

Furfural (2-furaldehyde) is a selective extractive solvent in the lubricating oil sector of the petroleum industry. It is also used to produce furfuryl alcohol, which in turn is used to produce "furan resins," important in the metal casting industry to form cores and molds. Smaller quantities of these resins are used to produce corrosion resistant grouts, mortars, joints, and valves. Some furfural is converted to tetrahydrofuran, a solvent in the resins and plastics industry. Furfural is produced by acid hydrolyzing the pentosans in agricultural by-products.

Lactic Acid

Lactic acid (2-hydroxypropanoic acid) is a common metabolite in animals, plants, and bacteria. It is widely used in the food industry as a preservative and flavoring. Esterification with ethanol produces ethyl lactate, a solvent that also serves as a chemical intermediate. Lactate can be converted to polylactic acid (PLA) resin and used in the production of biodegradable plastic that has properties competitive with plastics made from polyethylene and polystyrene. Although manufacturing processes have been developed for producing lactic acid from either fossil resources or biorenewable resources, economics are increasingly favoring the use of biorenewable resources. In this case, lactic acid is produced by fermentation of glucose.

Gluconic Acid

Gluconic acid (2,3,4,5,6-pentahydroxyhexanoic acid) is converted to the chelating agent sodium gluconate, useful in removing metal ions, especially calcium, magnesium, iron, and aluminum. As such, it finds application in metal cleaning as well as equipment cleaning in the dairy and food service industries. It also has miscellaneous applications in textile, pharmaceutical, and health food industries. Gluconic acid is produced in almost quantitative yields by the oxidation of glucose in air or oxygen.

Xylitol

Xylitol is a crystalline alcohol ($C_5H_{12}O_5$) derived from the five-carbon sugar xylose. Like sorbitol and mannitol, it is used as a sweetener and humidity control agent, although it has a relatively small market. Xylitol is produced in very high yield by the catalytic reduction of xylose in the presence of hydrogen.

Fermentative routes to xylitol production are under development. One strain of yeast, *Aureobasidium*, hydrolyzes corn fiber to xylose, while another strain, *Pichia guilliermondii*, ferments the xylose into xylitol.

Sorbitol

Sorbitol is a faintly sweet alcohol ($C_6H_{14}O_6$) derived from the catalytic hydrogenation of sucrose. Its use in toothpaste, cosmetics, and toiletries represents 32% of total domestic consumption of sorbitol. Its hygroscopic properties are employed as an emollient in creams and lotions, while its cooling, sweet taste makes it suitable for mouthwashes. Sorbitol is classified as a semi-natural, nutritive sweetener with 60% of the sweetness of sucrose; thus it is used as a confection in foods and snacks, especially mints and gums. It has applications as a bulking agent, peeling aid, and flavoring agent in the food industry. About 17% of sorbitol production is employed in the manufacture of vitamin C. Esterification of sorbitol results in a surfactant useful as a lubricant, softener, plasticizer, and anti-static agent. Sorbitol is made in very high yield by catalytic hydrogenation of isoglucose, which is produced by molybdate-catalyzed isomerization of starch-derived glucose.

Mannitol

Mannitol ($C_6H_{14}O_6$), an isomer of sorbitol, is also a slightly sweet alcohol. It finds similar applications as sorbitol, although it has a smaller market. Mannitol is a co-product of the production of sorbitol, representing about 30% of the yield from isoglucose.

Cellulose Derivatives

Cellulose can be manipulated to produce a variety of products. Most prominent is synthetic fibers of rayon, derived from regenerated cellulose. Rayon has found application in top-weight apparel, home furnishings, and non-woven materials. Cellulose acetate fibers, another product derived from cellulose, has wide application in the textile market for soft, dyeable apparel. Cellulose ester was the first thermoplastic, widely used to manufacture tools, toys, and automobile parts; however, this market has declined in the face of competition from petroleum-derived plastics. Similarly, the cellulose ether, known familiarly as cellophane, was widely used by the packing industry until less costly petroleum-based products were developed. Since cellophane is biodegradable its prospects might improve in the future. Most of these cellulose derivatives come from wood pulp.

5.4.3 Chemicals with Potential for Production from Biorenewable Resources

Levulinic Acid

Levulinic acid (4-oxopentanoic acid) has not been commercially exploited, but it may prove useful in the production of antifreeze, fuel additives, plasticizers, synthetic resins, and hydraulic brake fluids. Hydrogenation of levulinic acid yields valeric g-lactone, an excellent solvent. Levulinic acid is produced by high temperature acid hydrolysis of carbohydrates.

Levoglucosan

Levoglucosan is an anhydrosugar derived from cellulose or starch. It is a chiral compound with potential as a building block for the production of pharmaceuticals, insecticides, new polymers, and sugar alcohols. It is currently an expensive specialty chemical, but process improvements under development could drastically reduce its cost and open the way for commercial applications. Levoglucosan is produced by the fast pyrolysis of acid-treated lignocelluose with yields in the range of 20–30%. Separation techniques to yield a high purity product still need to be developed.

Hydroxyacetaldehyde

Hydroxyacetaldehyde is a carbonyl compound not currently produced commercially. However, it is a potential replacement for another carbonyl compound glyoxal (1,2-ethanedione), an important cross-linking agent in the manufacture of resins and is currently produced from fossil resources.

Hydroxyacetaldehyde is a co-product in the pyrolysis of lignocellulose, with yields of about 10%.

Starch Plastics

The earliest commercial "starch plastics" were starch-filled polyethylene, a composite derived from both fossil and biorenewable resources. Although they were "bio-disintegratable," with the starch breaking down to leave a recalcitrant polyethylene residue, they were not biodegradable in the sense of completely breaking down in relatively short time frames of a few weeks or months.

More advanced starch plastics chemically modify starch to produce polymeric material that completely degrades. For example, modification of hydroxyl groups during starch esterification yields starch esters that are thermoplastic and water resistance. Starch ester resin reinforced with biofibers has properties comparable to general-purpose polystyrene. Formulated with plasticizers and other additives, starch ester yields resins suitable for injection-molded products.

Acetylated Wood

Acetylated wood undergoes a chemical treatment that improves dimensional stability (that is, reduces swelling and shrinkage with moisture content) and resistance to biological degradation. The chemical treatment is a process of esterification and can be accomplished several different ways. Most commonly, acetic anhydride (ethanoic anhydride) is used to prepare acetylated wood. This process produces acetic acid as a by-product and unreacted acetic anhydride, which can be recycled to the acetylation reactor.

Polyhydroxybutyrate/polyhydroxyvalerate

Polyhydroxybutyrate (PHB) and its copolymers with polyhydroxyvalerate (PHV) are melt-processable, semi-crystalline thermoplastic polyesters that are stable under everyday conditions but degrade slowly in the body, when composted, or landfilled. The PHB homopolymer is stiff and rather brittle with mechanical properties resembling those of polystyrene, although it is less brittle and more temperature resistant. Copolymers are preferred for general purposes while the homopolymer is expected to have applications in the medical and biological fields. These polymers can be prepared microbiologically from glucose or syngas. Their production in transgenic plants has also been demonstrated.

5.5 Fibers

Plant fibers can be used in the manufacture of textiles, paper products, and composite materials. Cotton fibers are widely employed in the manufacture of textiles, while hardwood fibers dominate the markets for paper products and composite materials. Historically, non-woody plant fibers, such as wheat straw and corn stover, have also found application in the production of paper and composite materials, especially during times of wood scarcity. But the relatively higher cost of collecting and storing non-woody plant fibers and their slightly inferior properties has favored the use of wood fibers.

Worldwide, non-woody plant fiber accounts for almost 12% of pulping capacity, with wheat straw accounting for 40% of this capacity. Although small compared to wood fiber pulping, it represents a near doubling of the market share since 1970. Much of this capacity exists in developing countries: China, India, Pakistan, and Mexico, among others. The combination of increasing standard of living and growing shortage of wood in these countries is expected to further boost the potential of non-woody fiber in the pulp and paper industry.

A composite is a material consisting of fibers reinforced by an adhesive matrix. In general, composites contain a large volume fraction of adhesive to avoid the occurrence of voids, which markedly degrades the strength of composite materials. However, because plant fiber composites have been targeted at large volume markets, they rarely contain more than 3–12% adhesive to reduce the cost of the final product. This can be done because the polysaccharide and phenolic polymers from which plant fibers are constructed are very reactive and more readily bond to adhesives than do synthetic fibers. The inevitable presence of voids means that these products will be relegated to relatively low-value applications unless developers are able to capitalize on the unique properties imparted by the use of plant fibers in the composites.

Replacement of synthetic fibers by plant fibers is one area under development. As shown in Table 5.5, plant fibers can be competitive in terms of weight, specific tensile strength, and cost. Plant fibers also provide environmental advantages: less energy is consumed in producing these fibers, and the resulting composites can be burned for energy at the end of the life of the product.

One of the most important features of fibers is their aspect ratio; that is, the ratio of fiber length to diameter. Plant fibers have two distinct aspect ratios: one for individual fibers and another for fiber bundles that occur naturally in plant materials and can be extracted intact from certain non-wood plants, such as sisal, hemp, and flax. Most individual fibers have aspect ratios in the range of 100–200 with a few notable exceptions: hemp at 550, flax at 1500, cotton at 2000, and ramie at

Table 5.5. Comparison of properties of plant fibers to synthetic fibers

	Specific Gravity	Specific Tensile Strength (GPa)	Cost ($/ton)
Plant fibers	0.6–1.2	1.60–2.95	200–1000
Glass	2.6	1.35	1500–2000
Kevlar	1.4	2.71	4000–6000
Carbon	1.8	1.71	8000

Source: A.J. Bolton, 1994, "Natural Fibres for Plastic Reinforcement," *Materials Technology* 9: 12-20.

4000. The aspect ratio of fiber bundles varies from about 100 to over 4000, depending on the success of extracting the bundles from plant tissue. Short plant fibers are attractive as reinforcement in molding compounds and injection moldings, while longer fibers find application where anisotropy is an important property of the composites.

Two other properties of fibers will be mentioned that favor plant fibers in certain applications. Long fibers or fiber bundles can be aligned in a process known as *carding* in the textile industry. This ability to control fiber orientation allows anisotropic properties of strength and stiffness to be exploited. The other property, characteristic of long plant fibers, is known as *drape*: the ability of fabrics to flow over and follow complex shapes. This property is important in the molding of composites into complex shapes.

Further Reading

Fired Process Equipment

Perry, R. H. and C. H. Chilton. 1973. *Chemical Engineer's Handbook*, Fifth Edition. New York: McGraw-Hill.

Power Equipment

Culp, A. W. 1991. *Principles of Energy Conversion*, Second Edition. New York: McGraw-Hill.
Fuel Cell Handbook, Fifth Edition. Oct. 2000. (CD-ROM). Morgantown, WV/Pittsburgh, PA: US Department of Energy National Energy Technology Laboratory.
Moran, M. and Shapiro, H. 1999. *Fundamentals of Engineering Thermodynamics*, 4th Ed. New York: Wiley.

Transportation Fuels

Borman, G. L. and K. W. Ragland. 1998. *Combustion Engineering*. New York: McGraw-Hill.
Klass, D. L. 1998. "Synthetic oxygenated liquid fuels." Chap 11 in *Biomass for Renewable Energy, Fuels, and Chemicals*. San Diego: Academic Press.

Chemicals

Bozell, J. J. and R. Landucci, eds. 1993. *Alternative Feedstocks Program Technical and Economic Assessment: Thermal/Chemical and Bioprocessing Components*. U.S. Department of Energy Report.

Klass, D. L. 1998. "Organic commodity chemicals from biomass." Chap 13 in *Biomass for Renewable Energy, Fuels, and Chemicals*. San Diego: Academic Press.

Rowell, R. M., T. P. Schultz and R. Narayan, eds. 1992. *Emerging Technologies for Material and Chemicals from Biomass*. ACS Symp. Series 476. Washington, D.C.: American Chemical Society.

Stevens, E. S. 2002. *Green Plastics: An Introduction to the New Science of Biodegradable Plastics*. Princeton, NJ: Princeton University Press.

Fibers

Bolton, J. 1997. *The Burgess-Lane Memorial Lectureship in Forestry*. (Available on the World Wide Web at www.forestry.ubc.ca/burgess/bolton.html)

Rowell, R. M. 1992. "Opportunities for Lignocellulosic Materials and Composites." In *Emerging Technologies for Materials and Chemicals from Biomass*. R. M. Rowell, T. P. Schultz and R. Narayan, eds. ACS Symposium Series 476. Washington, D.C.: American Chemical Society.

Rowell, R. M., R. A. Young and J. R. Rowell, eds. 1997. *Paper and Composites from Agro-Based Resources*. CRC Press.

Problems

5.1 A 50 MW power plant, expected to operate at 42% thermodynamic efficiency, will use biomass for fuel. Determine how many hectares of land will be required to run it for the following biomass fuels:

(a) Hybrid poplar grown in North America

(b) Bagasse grown in Hawaii

5.2 Calculate the annual fuel costs for 10 MW power plants based on the following cycles. The fuel is hybrid poplar at $48/ton.

(a) Conventional steam power plant with a heat rate of 14,000 Btu/kWh.

(b) Integrated gasification/combined cycle (IGCC) based on gas turbines with overall thermodynamic efficiency of 50% based on gaseous fuel. However, the gasifier is only 85% energy efficient in converting the solid biomass fuel into gaseous fuel.

(c) Integrated gasification/combined cycle based on molten carbonate fuel cell.

5.3 For each of the following biobased transportation fuels, indicate the kind of engines in which they are targeted for use, their properties relative to the fuels they are displacing, and the major feedstock currently used to produce them in the United States.

(a) Ethanol

(b) Methyl esters

5.4 List at least three factors that complicate the use of biorenewable resources for the production of organic chemicals currently derived from petroleum.

5.5 List the seven basic building blocks of the "petroleum economy." Propose a short list of basic building blocks for a "bioeconomy."

CHAPTER 6

Conversion of Biorenewable Resources into Heat and Power

6.1 Introduction

Three approaches to converting biorenewable resources into heat and power are considered in this chapter: direct combustion, thermal gasification, and anaerobic digestion. Direct combustion of solid biomass yields hot flue gas that can be used in either the direct-fired or indirect-fired process heaters described in Chapter 5. Direct combustion can provide heat to Stirling engines or Rankine steam cycles for the generation of electric power. Both thermal gasification and anaerobic digestion yield flammable gas suitable for combustion in process heaters. This gas can also be conditioned for firing in internal combustion engines, gas turbines, and fuel cells. Thermal gasification is most suitable for low moisture biomass while anaerobic digestion can convert high moisture biomass to gaseous fuel.

6.2 Direct Combustion

Combustion is the rapid oxidation of fuel to obtain energy in the form of heat. Since biomass fuels are primarily composed of carbon, hydrogen, and oxygen, the main oxidation products are carbon dioxide and water. Flame temperatures can exceed 2000°C depending on the heating value and moisture content of the fuel, the amount of air used to burn the fuel, and the construction of the furnace.

6.2.1 Fundamentals of Combustion

Solid-fuel combustion consists of four steps, illustrated in Fig. 6.1: heating and drying, pyrolysis, flaming combustion, and char combustion. Heating and drying of the fuel particle is normally not accompanied by chemical reaction. Water is driven from the fuel particle as the thermal front advances into the interior of the particle. As long as water remains, the temperature of the particle cannot rise high enough to initiate pyrolysis, the second step in solid fuel combustion.

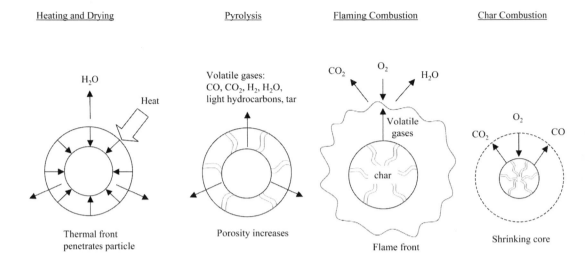

FIG. 6.1. Processes of solid fuel combustion.

Pyrolysis is a complicated series of thermally driven chemical reactions that decompose organic compounds in the fuel. Pyrolysis proceeds at relatively low temperatures, depending on the type of plant material. Hemicellulose begins to pyrolyze at temperatures between 225 and 325°C and lignin pyrolysis is initiated between 250 and 500°C.

The resulting decomposition yields a large variety of volatile organic and inorganic compounds, the types and amounts vary with the type and heating rate of the fuel. Pyrolysis products include carbon monoxide (CO), carbon dioxide (CO_2), methane (CH_4), and high molecular weight compounds that condense to a tarry liquid if cooled before they are able to burn. Fine droplets of these condensable compounds represent much of the smoke associated with smoldering fires. Pyrolysis follows the thermal front through the particle, releasing volatile compounds and leaving behind pores that penetrate to the surface of the particle.

Pyrolysis is very rapid compared to the overall burning process. It may be as short as a second for small particles of fuel, but can extend to many minutes in wood logs. Although the net result of combustion is oxidation of fuel molecules and the release of heat, neither of these processes occurs to a significant extent during pyrolysis. Indeed, if pyrolysis is to proceed at all heat must be added to the fuel. Oxygen is excluded from the pyrolysis zone by the large gaseous outflow of pyrolysis products from the surface of the fuel particle. Only after pyrolysis gases escape the particle and diffuse into the surrounding air are they able to burn. Upon completion of pyrolysis, a porous carbonaceous residue known as char remains.

Both the volatile gases and the char resulting from pyrolysis can be oxidized if sufficient oxygen is available to them. Oxidation of the volatile gases above the solid fuel results in *flaming combustion*. The ultimate products of volatile combustion are CO_2 and H_2O, although a variety of intermediate chemical compounds can exist in the flame, including CO, condensable organic compounds, and long chains of carbon known as soot. Indeed, hot, glowing soot is responsible for the familiar orange color of wood fires.

Combustion intermediates will be consumed in the flame if sufficient temperature, turbulence, and time are allowed. High combustion temperature assures that chemical reactions will proceed at high rates. Turbulent or vigorous mixing of air with the fuel makes certain that every fuel molecule comes into contact with oxygen molecules. Long residence times for fuel in a combustor allow the fuel to be completely consumed. In the absence of good combustion conditions, a variety of noxious organic compounds can survive the combustion process including CO, soot, polycyclic aromatic hydrocarbons (PAH), and the particularly toxic families of chlorinated hydrocarbons known as furans and dioxins. In some cases, a poorly operated combustor can produce pollutants from relatively benign fuel molecules.

The next step in combustion of solid fuels is *gas-solid reactions* of char, known as *char combustion* or *glowing combustion,* familiar as red-hot embers in a fire. Char is primarily carbon with a small amount of mineral matter interspersed. Char oxidation is controlled by mass transfer of oxygen to the char surface rather than by chemical kinetics, and is very fast at the elevated temperatures of combustion. Depending on the porosity and reactivity of the char and the combustion temperature, oxygen may react with char at the surface of the particle, or it may penetrate into the pores before oxidizing char inside the particle. The former situation results in a steadily shrinking core of char, whereas the latter situation produces a constant diameter particle of increasing porosity. Both CO and CO_2 can form at or near the surface of burning char:

$$C + \tfrac{1}{2}O_2 \rightarrow CO \tag{6.1}$$
$$CO + \tfrac{1}{2}O_2 \rightarrow CO_2 \tag{6.2}$$

These gases escape the immediate vicinity of the char particle where CO is oxidized to CO_2 if sufficient oxygen and temperature are available; otherwise it appears in the flue gas as a pollutant.

6.2.2 Combustion Equipment

A *combustor* is a device that converts the chemical energy of fuels into high temperature exhaust gases. Heat from the high temperature gases can be employed in a variety of applications, including space heating, drying, and power generation. However, with the exception of kilns used by the cement industry, most solid-fuel combustors today are designed to produce either low-pressure steam for process heat or high-pressure steam for power generation. Combustors integrated with steam-raising equipment are called *boilers*. In some boiler designs, distinct sections exist for combustion, high temperature heat transfer, and moderate temperature heat transfer: these are called the *furnace, radiative section,* and *convective section* of the boiler, respectively. In other designs, no clear separation between the processes of combustion and heat transfer exists.

Solid-fuel combustors, illustrated in Fig. 6.2, can generally be categorized as grate-fired systems, suspension burners, or fluidized beds. *Grate-fired systems* were the first burner systems to be developed, evolving during the late nineteenth and early twentieth centuries into a variety of automated systems. The most common system is the *spreader-stoker,* consisting of a fuel feeder that mechanically or pneumatically flings fuel onto a moving grate where the fuel burns. Much of the ash falls off the end of the moving grate although some fly ash appears in the flue gas. Grate systems rarely achieve combustion efficiencies exceeding 90%.

Suspension burners were introduced in the 1920s as a means of efficiently burning large quantities of coal pulverized to less than 50-μ particle sizes. Suspension burners suspend the fuel as fine powder in a stream of vertically rising air. The fuel burns in a fireball and radiates heat to tubes that contain water to be converted into steam. Suspension burners, also know as *pulverized coal (PC)*

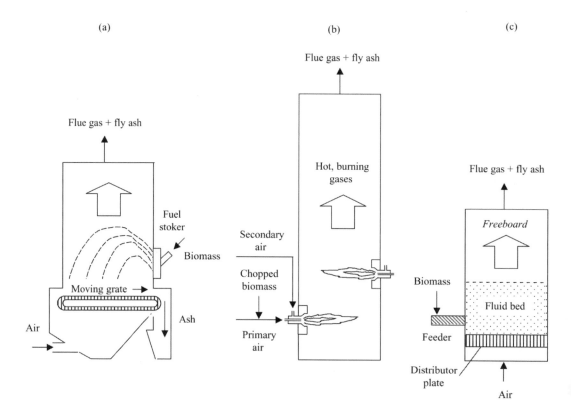

FIG. 6.2. Common types of combustors: (a) grate-fired, (b) suspension, (c) fluidized bed.

boilers, have dominated the U.S. power industry since World War II because of their high volumetric heat release rates and their ability to achieve combustion efficiencies, often exceeding 99%. However, they are not well suited to burning coarse particles of biomass fuel, and they are notorious generators of nitrogen oxides. Biomass is fed from a bunker through pulverizers designed to reduce fuel particle size enough to burn in suspension. The fuel particles are suspended in the primary airflow and fed to the furnace section of the boiler through burner ports where the fuel burns as a rising fireball. Secondary air injected into the boiler helps complete the combustion process. Heat is absorbed as the hot exhaust passes by steam tubes arrayed in banks of heat exchangers (waterwall, superheaters, and economizer) before the exhaust exits through a baghouse designed to capture ash released from the fuel. Steam produced in the boiler is used in a Rankine power cycle.

Fluidized bed combustors are a recent innovation in boiler design. Air injected into the bottom of the boiler suspends a bed of sand or other granular refractory material producing a turbulent mixture of air and sand. The high rates of heat and mass transfer in this environment are ideal for efficiently burning a variety of fuels. Furthermore, the large thermal mass of the sand bed allows the unit to be operated as low as 850°C, which lowers the emission of nitrogen compounds. A commercial market for fluidized bed boilers developed during the 1980s, especially for industrial applications.

Recently, a *whole-tree burner* has been proposed for electrical utility steam-raising. The advantage of this approach is minimization of field processing and handling, and the elimination of fuel chipping. Whole-tree burning can save about 35% of the cost of harvesting and handling woody fuels.

A variety of biomass materials have proven suitable for direct combustion including whole trees, wood chips, forestry residue, agricultural residue, pulp and paper refuse, food processing wastes, municipal solid waste, and straws and grasses. Wood and wood wastes dominate the biomass-to-power market in the U.S. and much of the rest of the world. However, a variety of agricultural wastes have been utilized through direct combustion in California, and Denmark has a number of commercial plants that burn straw.

Direct combustion has the advantage that it employs commercially well-developed technology. There are a number of vendors who supply turnkey systems, and considerable operating experience exists both in the United States and abroad. However, there are three prominent disadvantages with direct firing. These include penalties associated with burning high-moisture fuels, agglomeration and ash fouling due to alkali compounds in biomass, and relatively low thermodynamic efficiencies for steam power plants of the size appropriate to biomass power.

The moisture in biomass degrades boiler performance for two reasons. First, the energy required to vaporize fuel moisture can only be recovered if exhaust gases are cooled sufficiently to condense the water vapor in the gas. Although this procedure would result in very high thermal efficiencies for a boiler burning even very high-moisture fuels, it is not commonly done in practice because the condensate is often corrosive to boiler tubes. Accordingly, most direct-combustion systems are penalized by moisture in the fuel. Second, high-moisture fuels simply do not burn well because the process of fuel drying suppresses fuel temperatures to below those required for ignition. For this reason water contents exceeding 30% are unacceptable in most boilers. Fluidized-bed combustors, however, can accept biomass with moisture content as high as 50% because of the enormous thermal mass associated with the hot bed material (although even fluidized beds are penalized in thermal efficiency by the presence of this moisture). Depending on moisture content, as much as 15% of the heating value of biomass is required to dry the fuel. Obviously, field drying of biomass is desirable to reduce both transportation costs and heating penalties within the boiler.

Alkali in biomass fuels presents a difficult problem for direct combustion systems. Compounds of alkali metals, such as potassium and sodium salts, are common in rapidly growing plant tissues. Annual biomass crops contain large quantities of alkali while the old-growth parts of perennial biomass contain relatively small quantities of alkali. These alkali compounds appear as oxides in the residue left after combustion of volatiles and char. Alkali vapors may combine with sulfur and silica to form low-melting-point compounds. These sticky compounds bind ash particles to fuel grates and heat exchanger surfaces. Boiler performance degrades as airflow and heat transfer are restricted by ash deposits. Boilers that fire straw, which contains both high silica and alkali, have experienced serious problems in slagging and fouling, as well as high temperature corrosion, unless steam temperatures are kept relatively low (less than 500°C). (*Slagging* is the partial or complete melting of ash, whereas *fouling* is the accumulation of sticky ash particles on heat exchange surfaces.)

As an alternative to completely replacing coal with biomass fuel in a boiler, mixtures of biomass and coal can be burned together in a process known as *co-firing*. Co-firing offers several advantages for industrial boilers. Industries that generate large quantities of biomass wastes, such as lumber mills or pulp and paper companies, can use co-firing as an alternative to costly landfilling of wastes. Federal regulations also make co-firing attractive. The New Source Performance Standards, which limit particulate emissions from large coal-fired industrial boilers to 0.05 pounds per million British thermal units (lb/MMBtu), double this allowance in co-fired boilers when the capacity factor for biomass exceeds 10%. Adopting co-firing is a good option for companies that are slightly out of compliance with their coal-fired boilers. Similarly, a relatively simple way for a company to reduce sulfur emissions from its boilers is to co-fire with biomass, which contains much less sulfur than coal. Co-firing capability also provides fuel flexibility and reduces ash-fouling problems associated with using only biomass as fuel.

The principal disadvantages of co-firing relate to the characteristics of biomass fuels. Because of the lower energy density and higher moisture content of biomass, the steam generating capacity of co-fired boilers is often reduced. Also, the elemental composition of fly ash from biomass is distinct from that of coal fly ash.

Utilities are concerned that co-mingled biomass and coal ash will not meet the American Society for Testing and Materials (ASTM) definition of fly ash that is acceptable for concrete admixtures, thus eliminating an important market for this combustion by-product.

It is generally recommended that total fuel alkali content, in single or mixed fuels, be limited to less than 0.17–0.34 kilograms per gigajoule (kg/GJ) (0.4–0.8 lb/MMBtu), which translates to only 5–15% co-firing of biomass with coal. Also, furnace temperatures should be kept below 980°C to help prevent the buildup of alkali-containing mineral combinations known as *eutectics*. At higher temperatures, molten eutectics bind dirt and other particulate matter to form slag and fouling deposits.

The best wood-fired power plants, which are typically 20–100 MW in capacity, have heat rates exceeding 12,500 Btu/kWh. In contrast, large coal-fired power plants have heat rates of only 10,250 Btu/kWh. The relatively low thermodynamic efficiency of steam power plants of the sizes that would be used in biomass power systems may ultimately limit the potential of direct combustion to convert biomass fuels to useful energy.

6.3 Thermal Gasification

Gasification is the high temperature (750–850°C) conversion of solid, carbonaceous fuels into flammable gas mixtures. This gas, sometimes known as producer gas, consists of carbon monoxide (CO), hydrogen (H_2), methane (CH_4), nitrogen (N_2), carbon dioxide (CO_2), and smaller quantities of higher hydrocarbons. The overall process is endothermic; that is, it requires a source of heat to drive the process. This heat may be provided externally from the gasifier or internally by burning part of the fuel entering the gasifier. Gasification should not be confused with anaerobic digestion, the microbial degradation of biomass to gas described in the next section.

Gasification was placed into commercial practice as early as 1812 when, in England, coal was converted to gas for illumination (known as manufactured gas or town gas). This technology was widely adopted in industrialized nations and was employed in the United States as late as the 1950s when interstate pipelines made inexpensive supplies of natural gas available. In some places, such as China, gas is still manufactured from coal.

The high volatile content of biomass (70–90 wt-%) compared to coal (typically 30–40 wt-%), and the high reactivity of its char, make biomass an ideal gasification fuel. However, issues of cost and convenience of biomass gasification have limited its applications to special situations and niche markets. For example, in response to petroleum shortages during World War II, over one million small-scale wood gasifiers were built to supply low-enthalpy gas for automobiles and steam boilers. These were abandoned soon after the war.

Not only can producer gas be used for generation of heat and power, but it can also serve as feedstock for production of liquid fuels and chemicals as described in

Chapter 7. Because of this flexibility of application, gasification has been proposed as the basis for "energy refineries" that would provide a variety of energy and chemical products, including electricity and transportation fuels.

6.3.1 Fundamentals of Gasification

Figure 6.3 illustrates the several steps of gasification: heating and drying, pyrolysis, solid-gas reactions that consume char, and gas-phase reactions that adjust the final chemical composition of the producer gas. Drying and pyrolysis are similar to the corresponding processes for direct combustion described in the previous section. Pyrolysis begins between 300 and 400°C and produces the intermediate products of char, gases (mainly CO, CO_2, H_2, and light hydrocarbons) and condensable vapor (including water, methanol, acetic acid, acetone, and heavy hydrocarbons). The distribution of these products depends on the chemical composition of the fuel and the heating rate and temperature achieved in the reactor. However, the total pyrolysis yield of pyrolysis products, and the amount of char residue, can be roughly estimated from the proximate analysis of the fuel. The fuel's volatile matter roughly corresponds to the pyrolysis yield, while the combination of fixed carbon and ash content can be used to estimate the char yield.

Heating and drying are *endothermic processes,* and the heat required to drive them can be supplied by an external source in a process known as *indirectly heated gasification.* More typically, a small amount of air or oxygen (typically not more than 25% of the stoichiometric requirement for complete combustion of the fuel) is admitted

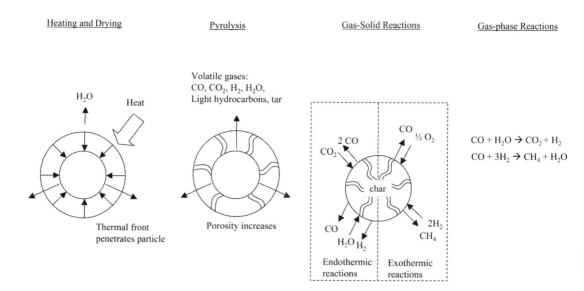

FIG. 6.3. Processes of thermal gasification.

for the purpose of partial oxidation. This releases sufficient heat for drying and pyrolysis, as well as for the subsequent endothermic chemical reactions described below.

The third step of gasification is *gas-solid reactions*. These reactions convert solid carbon into gaseous CO, H_2, and CH_4:

Carbon-oxygen
reaction: $C + \frac{1}{2}O_2 \leftrightarrow CO$ $\Delta H_R = -110.5 \text{ MJ/kmol}$ (6.3)

Boudouard
reaction: $C + CO_2 \leftrightarrow 2CO$ $\Delta H_R = 172.4 \text{ MJ/kmol}$ (6.4)

Carbon-water
reaction: $C + H_2O \leftrightarrow H_2 + CO$ $\Delta H_R = 131.3 \text{ MJ/kmol}$ (6.5)

Hydrogenation
reaction: $C + 2H_2 \leftrightarrow CH_4$ $\Delta H_R = -74.8 \text{ MJ/kmol}$ (6.6)

The first of these, known as the *carbon-oxygen reaction*, is strongly exothermic and is important in supplying the energy requirements for drying, pyrolysis, and endothermic gas-solid reactions. The *hydrogenation reaction* also contributes to the energy requirements of the gasifier, although significantly more char reacts with oxygen than hydrogen in the typical air-blown gasifier.

The fourth step of gasification is *gas-phase reactions*, which determine the final mix of gaseous products:

Water-gas
shift reaction: $CO + H_2O \leftrightarrow H_2 + CO_2$ $\Delta H_R = -41.1 \text{ MJ/kmol}$ (6.7)

Methanation: $CO + 3H_2 \leftrightarrow CH_4 + H_2O$ $\Delta H_R = -206.1 \text{ MJ/kmol}$ (6.8)

The final gas composition is strongly dependent on the amount of oxygen and steam admitted to the reactor as well as the time and temperature of reaction. For sufficiently long reaction times, chemical equilibrium is attained and the products are essentially limited to the light gases: CO, CO_2, H_2, and CH_4 (and N_2 if air is used as a source of oxygen). Analysis of the chemical thermodynamics of these six gasification reactions reveals that low temperatures and high pressures favor the formation of CH_4, whereas high temperatures and low pressures favor the formation of H_2 and CO.

Often gasifier temperatures and reaction times are not sufficient to attain chemical equilibrium, and the producer gas contains various amounts of light hydrocarbons such as C_2H_2 and C_2H_4 as well as up to 10 wt-% heavy hydrocarbons that condense to a black, viscous liquid known as "*tar*." This latter product is undesirable as it can block valves and filters and interferes with downstream conversion processes. Steam injection and addition of catalysts to the reactor are sometimes used to shift products toward lower molecular weight compounds.

6.3.2 Gasification Equipment

Gasifiers are generally classified according to the method of contacting fuel and gas. The three main types of interest to biomass gasification are updraft (counter-current), downdraft (concurrent), and fluidized bed. These are illustrated in Fig. 6.4, while their performance characteristics are summarized in Table 6.1.

Updraft gasifiers, illustrated in Fig. 6.4a, are the simplest as well as the first type of gasifier developed. They were a natural evolution from charcoal kilns, which yield smoky yet flammable gas as a waste product, and blast furnaces, which generate product gas that reduces ore to metallic iron. Updraft gasifiers are little more than grate furnaces with chipped fuel admitted from above and insufficient air for complete combustion entering from below. Above the grate, where air first contacts the fuel, combustion occurs and very high temperatures are produced. Although the gas flow is depleted of oxygen higher in the fuel bed, hot H_2O and CO_2 from combustion near the grate reduce char to H_2 and CO. These reactions cool the gas, but temperatures are still high enough to heat, dry, and pyrolyze the fuel moving down toward the grate. Of course, pyrolysis releases both condensable and non-condensable gases, and the producer gas leaving an updraft gasifier con-

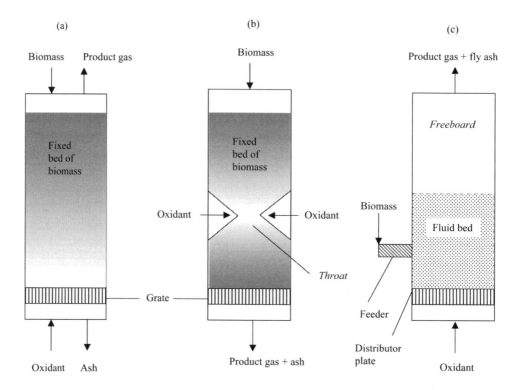

FIG. 6.4. Common types of biomass gasifiers: (a) updraft, (b) downdraft, (c) fluidized bed.

Table 6.1. Producer gas composition from various kinds of gasifiers

Gasifier Type	Gaseous Constituents (vol.% dry)					Energy Content	Gas Quality	
	H_2	CO	CO_2	CH_4	N_2	HHV (MJ/m^3)	Tars	Dust
Air-blown updraft	11	24	9	3	53	5.5	High (~10 g/m^3)	Low
Air-blown downdraft	17	21	13	1	48	5.7	Low (~1 g/m^3)	Medium
Air-blown fluidized bed	9	14	20	7	50	5.4	Medium (~10 g/m^3)	High
Oxygen-blown downdraft	32	48	15	2	3	10.4	Low (~1 g/m^3)	Low
Indirectly heated fluidized bed	31	48	0	21	0	17.4	Medium (~10 g/m^3)	High

Source: Adapted from A. V. Bridgwater, "The technical and economic feasiblity of biomass gasification for power generation," *Fuel 74*, 631-653, 1995; T.A. Milne, N. Abatzoglou, R. J. Evans, "Biomass Gasifier 'Tars': Their nature, formation, and conversion," *National Renewable Energy Laboratory Report*, NREL/TP-570-25357, 1998.

tains large quantities of tars, on the order of 50 g/m^3. As a result, updraft gasifiers are generally not strong candidates for biomass energy applications.

In *downdraft gasifiers*, fuel and gas move in the same direction. Downdraft gasifiers appear to have been developed near the end of the nineteenth century after the introduction of induced-draft fans allowed air to be drawn downward through a gasifier in the same direction as the gravity-fed fuel. As shown in Fig. 6.4b, contemporary designs usually add an arrangement of air nozzles, or tuyeres, that admit air or oxygen directly into a region known as the throat where combustion forms a bed of hot char. This design assures that condensable gases released during pyrolysis are forced to flow through the hot char bed, thus *cracking*, or thermally decomposing, tars. The producer gas is relatively free of tar (less than 1 g/m^3), making it a satisfactory fuel for engines. A disadvantage is the need for tightly controlled fuel properties (particles sized to between 1 and 30 cm, low ash content, and moisture less than 30%). Another disadvantage is a tendency for slagging or sintering of ash in the concentrated oxidation zone. Rotating ash grates or similar mechanisms can solve this problem. Furthermore, incorporation of a throat region limits the maximum size of downdraft gasifiers to about 400 kg/hr.

In a *fluidized bed gasifier*, illustrated in Fig. 6.4c, a gas stream passes vertically upward through a bed of inert particulate material to form a turbulent mixture of gas and solid. Fuel is added at such a rate that it is only a few percent by weight of the bed inventory. Unlike the updraft and downdraft gasifiers, no segregated regions of combustion, pyrolysis, and tar cracking exist. The violent stirring action makes the bed uniform in temperature and composition with the result that gasification occurs simultaneously at all locations in the bed. Typically a fluidized gasifier operates in the range of 700–850°C. By injecting fuel in the base of the bed, much of the tar can be cracked within the fluidized bed. However, a large insulated space above the bed, known as the freeboard, is usually included to promote additional tar cracking as well as more complete conversion of char. Nevertheless, tar production is intermediate between updraft and downdraft gasifiers (about 10 g/m^3). Fluidized beds are attractive for biomass gasification. They are able to process a wide variety of fuels including those of high moisture and small size. They are easily scaled to large sizes suitable for electric power production. Disadvantages of fluidized beds include relatively high power consumption to move gas through the bed; high exit gas temperatures, which complicate efficient energy recovery; and relatively high particulate burdens in the gas due to the abrasive forces acting within the fluidized bed.

A fourth gasifier of questionable usefulness in biomass gasification employs finely pulverized fuel in an *entrained flow reactor*. This reactor was developed for steam-oxygen gasification of coal at temperatures of 1200–1500°C. These high temperatures assure excellent char conversion (approaching 100%) and low tar production. The technology is attractive for advanced coal power plants but is unlikely to be used for commercial biomass gasification. Processing of biomass material into finely divided particles required for entrained flow gasifiers is prohibitively expensive. Furthermore, the low energy density and high sintering potential of biomass makes it difficult to achieve the high temperatures characteristic of entrained flow gasifiers.

The fuel/air ratio is the single most important parameter for determining gasifier performance. Downdraft gasifiers can have better conversion efficiency and producer gas quality than fluidized bed gasifiers because they utilize a higher fuel/air ratio. Fluidized bed gasifiers, on the other hand, can generally handle a wider range of biomass feedstocks with higher moisture content.

Recent research in biomass gasification has focused on improving the heating value of product gas. Conventional gasification admits sufficient air or oxygen to the reactor to burn part of the fuel, thus releasing heat to support pyrolysis of the rest of the fuel. Gas produced in air-blown biomass gasifiers typically has heating value that is only 10–20% that of natural gas. This low heating value is largely the result of nitrogen diluting the fuel gas. Oxygen can be used as the fluidization agent, but high capital costs preclude this from consideration at the relatively small sizes envisioned for most biomass energy systems.

Indirectly heated gasification can improve gas heating value by physically separating combustion and pyrolysis. As a result, the products of combustion do not appear in the fuel gas. Higher heating values of 14,200 kJ/m^3 or higher are expected. Several schemes have been suggested for transporting heat from the combustion reactor to the pyrolysis reactor. These include transferring hot solids from the combustor to the pyrolyzer, transferring a chemically regenerative heat carrier between two reactors, transferring heat through a wall common to the reactors, and storing heat in high-temperature phase-change material.

Depending on the kind of gasifier, product gas can be contaminated by tar and particulate matter. Tar cannot be tolerated in many downstream applications, including internal combustion engines, fuel cells, and chemical synthesis reactors. Tar can deposit on surfaces in filters, heat exchangers, and engines where it reduces component performance and increases maintenance requirements. A well-designed gasifier has a high temperature zone, through which pyrolysis products pass, allowing tars to crack into the low molecular gases predicted by equilibrium theory. The production of particulate matter, a combination of ash and unreacted char from the biomass, varies considerably from one type of gasifier to another. Particulate matter can also deposit on critical power system components, degrading performance even at relatively low concentrations in some machinery.

Gas cleaning usually involves separate processes for removing solid particles and tar vapor or droplets. A gas cyclone operated at a temperature above the condensation temperature for the least volatile tar constituent can remove most ash and char particles larger than about 5 μm in diameter. If removal of finer particles is required additional filtration is required, such as ceramic-candle filters or moving-bed granular filters.

Two approaches are available for removing tar from product gas. Scrubbing the gas stream with a fine mist of water or oil removes both tar and particles from the gas stream. Gas scrubbers are widely used in the chemical industry and are relatively inexpensive. However, cooling the gas and removing organic compounds reduce both sensible energy and chemical energy, which decreases overall energy conversion efficiency. Gas scrubbing also produces a toxic stream of tar, which complicates waste disposal.

The other approach to removing tar converts it into low molecular weight compounds. This approach has the advantage of increasing the heating value of the producer gas while eliminating a waste disposal problem. This cracking of tar into CO and H_2 can be accomplished by raising the gas to temperatures exceeding 1000°C. Such high temperatures are not easily achieved in most biomass gasifiers. The use of catalysts allows tar cracking to occur at much lower temperatures, in the range of 600–800°C. The gas is passed through a packed bed reactor containing a catalyst with steam or oxygen sometimes used to promote reaction.

The thermodynamic efficiency of gasifiers is strongly dependent on the kind of gasifier and how the product gas is employed. Some high-temperature, high-pressure gasifiers are able to convert 90% of the chemical energy of solid fuels into chemical and sensible heat of the product gas. However, these high efficiencies come at high capital and operating costs. Most biomass gasifiers have conversion efficiencies ranging between 70 and 80%. In some applications, such as process heaters or driers, both the chemical and sensible heat of the product gas can be utilized. In many power applications, though, the hot product gas must be cooled before it is utilized; thus, the sensible heat of the gas is lost. In this case, "cold gas" efficiency can be as low as 50–60%. Whether the heat removed from the product gas can be recovered for other applications, like steam raising or fuel drying, ultimately determines which of these conversion efficiencies is most meaningful. Gas cleaning to remove tar and particulate matter also has a small negative impact on gasifier efficiency, since it removes flammable constituents from the gas (tar and char particles) and generally requires a small amount of energy to run pumps.

6.4 Anaerobic Digestion

Anaerobic digestion is the decomposition of organic wastes, including polysaccharides, proteins, and lipids, to gaseous fuel by bacteria in an oxygen-free environment. The process occurs in stages, each involving specific types of bacteria that successively break down the organic matter into simpler organic compounds. The desired product, known as biogas, is a mixture of CH_4, CO_2, and some trace gases. Due to the diverse hydrolytic enzymes arising from the mixed culture of bacteria involved in anaerobic digestion, this process is unique among microbial fermentations in its adaptability to numerous feedstocks.

Anaerobic digestion arises naturally in marshy ground and landfills. The English gave the name "will-o'-the-wisp" to the mysterious, ephemeral flames that appeared spontaneously in marshlands as a result of decaying vegetation. The origin of this flammable gas was understood by the middle of the seventeenth century but not exploited until the end of the following century when gas from sewage was used for street lighting in England and power in India. These early systems were difficult to operate and maintain, and hydrogen sulfide in the gas corroded pipes and engines. Today the technology is well developed for production of methane-rich gas from almost any kind of feedstock with the possible exception of lignins and keratins.

6.4.1 Fundamentals of Anaerobic Digestion

The complete anaerobic degradation of organic matter is a three-stage process, illustrated in Fig. 6.5, consisting of hydrolysis and fermentation, acidification, and methanogenesis involving diverse bacterial consortia. The first step involves *hydrolysis* of polysaccharides to oligosaccharides and monosaccharides, of proteins to peptides and amino acids, of triglycerides to fatty acids and glycerol, and of nucleic acids to heterocyclic nitrogen compounds, ribose, and inorganic phosphate. Hydrogen and carbon dioxide are also produced. Coliform bacteria, such as the pathogen *Escherichia coli* and pathogens in the genera *Salmonella* and *Shigella*, are common fermentative bacteria.

This step is followed by an *acidification* step of transitional digestion through the action of acetogenic (acid-forming) bacteria. Products of fermentation that are too complex for methanogenic (methane-forming) bacteria to consume are further degraded in this step to acetate, H_2, and CO_2. Traces of oxygen in the feedstock are

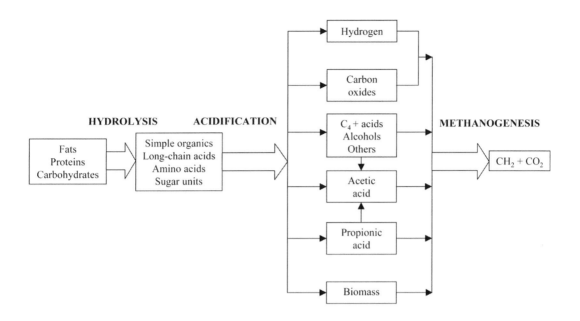

FIG. 6.5. Microbial phases in anaerobic digestion (adapted from D. L. Klass, 1998, *Biomass for Renewable Energy, Fuels and Chemicals*, San Diego: Academic Press).

consumed in this step, which benefits oxygen-sensitive, methanogenic bacteria. Furthermore, nutrients such as ammonium salts from degradation of proteins are released. The final step, *methanogenesis*, converts acetate to CH_4 and CO_2 by the action of methanogenic bacteria. Furthermore, H_2 and CO_2 released during acetogenesis are converted to additional methane. About 70% of the methane and carbon dioxide is produced from acetate and 30% is produced from carbon dioxide and hydrogen. The methanogenic bacteria lower the acidity of the fermentation broth, thus favoring acetogenesis as the acid-producing organisms are sensitive to high acidity.

Advantages of anaerobic digestion for processing biomass include the ability to use non-sterile reaction vessels; automatic product separation by outgassing; and relatively simple equipment and operations. The primary disadvantages are slow reaction rates and low methane yields. The breakdown of cellulose to sugars may require reaction times as long as a month to achieve high yields of methane. Some plant components are resistant to microbial degradation. Lignin, in particular, is only slowly converted because of its stable polyaromatic structure. Although the theoretical weight yield of methane from glucose is 27 wt-%, the complex of lignin and cellulose known as lignocellulose in plant materials results in substantially lower yields of methane than might otherwise be expected from cellulose. Another disadvantage is microbial reduction of sulfate and other sulfur compounds to hydrogen sulfide, a toxic, corrosive gas that complicates the use of biogas.

6.4.2 Anaerobic Digestion Equipment

All anaerobic digesters produce two effluent streams: biogas and sludge. Most digestion systems produce biogas that is between 55 and 75% methane by volume with the balance consisting of CO_2 along with traces of H_2S. The amount of biogas produced depends both on the composition of the feedstock and the design of the digester. Table 6.2 gives the range of gas yields expected for various feedstocks in terms of the volatile solids content of the material. *Volatile solids* are defined as that part of the dried feedstock material that is consumed upon firing in a muffle furnace at 550°C.

Anaerobic digesters can be broadly characterized as single-tank or two-tank reactors (also known as conventional or two-phase reactors, respectively). The *single-tank reactors*, illustrated in Fig. 6.6, can be operated in one of several modes: batch, intermittent, or continuous.

In the *batch mode*, the reactor is loaded with feedstock only once and microbial reactions are allowed to proceed to completion. Only gas is removed during the process. Batch reactors are inherently non-steady in their operation, which prevents optimum feedstock throughput and biogas yields. Thus, they are relegated to only the smallest digester systems.

The *intermittent mode* periodically adds additional feedstock as aqueous slurry while drawing off an equal volume of fermenter broth. The layers within the reactor are stratified into an active layer of digesting solids, an inactive layer of stabilized (indigestible) solids at the bottom, and a supernatant layer at the top. The surface of the supernatant layer is usually covered with a thin layer of scum. The challenge of this mode is retaining the sludge until it is fully digested. One way to improve efficiency is to encourage movement of feedstock through the active biosolids where the microorganisms reside. The intermittent mode requires retention times of 30–60 days and has loading rates of 0.5–1.5 kg per day of volatile solids per cubic meter capacity.

Table 6.2. Anaerobic digestion yields of biogas for various feedstocks

Feedstock	Biogas Yield (m³/kg of volatile solids)
Wastewater	0.093–0.31
Human sewage	0.37–0.93
Distillery waste	> 0.69

Source: Charles E. Wyman, 1994, Alternative Fuels from Biomass and Their Impact on Carbon Dioxide Accumulation, *Appl. Biochem. Biotech.* 45/46:897-915.

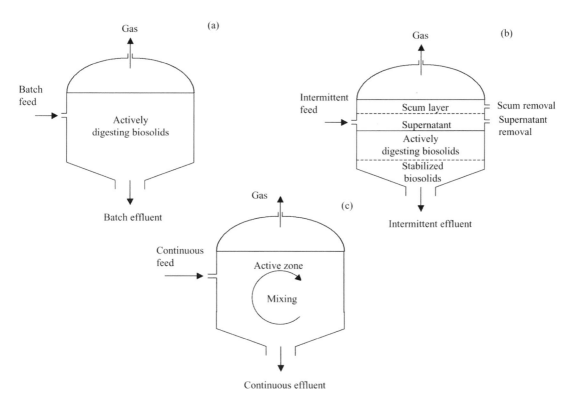

FIG. 6.6. Types of single-tank anaerobic digesters: (a) batch-fed, (b) intermittently fed, (c) continuously stirred and fed (adapted from D. L. Klass, 1998, *Biomass for Renewable Energy, Fuels and Chemicals*, San Diego: Academic Press).

Higher rate digestion can be achieved in *continuous mode* operation of single tank reactors. Mixing provides homogeneity. Retention times are 20 days or less, and loading rates are 1.6–6.4 kg per day of volatile solids per cubic meter capacity.

Two-tank reactors, illustrated in Fig. 6.7, were developed in response to the observation that methanogenesis appears to be the rate-limiting step in anaerobic digestion. Two-tank reactors physically separate the acid-formation and methane-formation phases of digestion so that each takes place under optimal conditions. Benefits of this technology are numerous: hydrolysis and acidification occur more quickly than in conventional systems; the common problems of foaming in single-tank systems are reduced by the biological destruction of biochemical foaming compounds before the feedstock reaches the methane-forming reactor; and the biogas produced is typically rich in methane.

Anaerobic digestion can be used for a wide range of feedstocks of medium-to-high moisture content. Sewage sludge is most commonly used as feedstock for

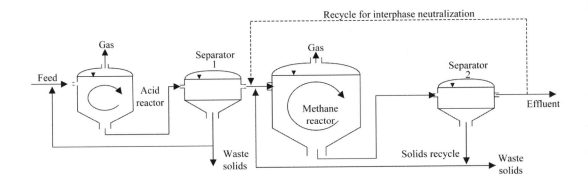

FIG. 6.7. Two-phase anaerobic digesters (adapted from D. L. Klass, 1998, *Biomass for Renewable Energy, Fuels and Chemicals*, San Diego: Academic Press).

anaerobic digesters, but municipal solid waste (MSW), food processing wastes, agricultural wastes, Napier grass, kelp, bagasse, and water hyacinth can also be digested. Digesters are designed to maintain optimal conditions for a specific type of waste. Waste pretreatment, heating, mixing, nutrient addition, specialized bacteria addition, and pH adjustment can be manipulated to control digester performance.

Biogas, once treated to remove H_2S, can substitute for natural gas in many applications, including stationary power generation. The CO_2 can be recovered and sold, and the digested solids are often marketable as fertilizer, animal bedding, and sometimes as animal feed. The digested solids have higher nutrient values per unit mass than the undigested feed solids and are effective slow-nitrogen-release fertilizers.

Further Reading

Combustion

Annamalai, K., M. Sami and M. Wooldridge. 2001. "Co-firing of Coal and Biomass Fuel Blends." *Progress in Energy & Combustion Science* 27:171–214.
Borman, G. L. and K. W. Ragland. 1998. *Combustion Engineering*. McGraw-Hill.
Jenkins, B. M., L. L. Baxter, T. R. Miles, Jr., and T. R. Miles, Sr. 1998. "Combustion Properties of Biomass." *Fuel Processing Technology* 54:17–46.

Thermal Gasification

Bridgwater, A. V. 1995. "The Technical and Economic Feasibility of Biomass Gasification for Power Generation." *Fuel* 74:631–653.
Reed, T., ed. 1981. *Biomass Gasification: Principles and Technology*. Park Ridge, N.J: Noyes Data Corp.

Anaerobic Digestion

Klass, D. L. 1998. "Microbial Conversion: Gasification." Chap. 12 in *Biomass for Renewable Energy, Fuels, and Chemicals*. San Diego: Academic Press.

Speece, R. E. 1996. *Anaerobic Biotechnology for Industrial Wastewaters*. Nashville, Tennessee: Archae Press.

Problems

6.1 The manager of a power plant proposes to co-fire switchgrass with coal as a means of reducing sulfur emissions from the plant. However, the plant engineer warns that alkali in the switchgrass can lead to ash fouling in the boiler. The raw coal, which contains virtually no alkali, has a heating value of 28 MJ/kg and sulfur content of 4 wt-%.

   (a) What is the maximum weight-percent of switchgrass that can be blended with the coal without producing ash fouling?

   (b) What is the expected sulfur emission rate (kg/GJ) for the fuel blend?

6.2 List the comparative advantages of combustion, gasification, and anaerobic digestion. Indicate an application most suitable for each one.

6.3 Describe the three kinds of gasifiers suitable for biomass gasification. Indicate an application most suitable for each one.

6.4 Show that the theoretical yield of methane from anaerobic digestion of cellulose is 27 wt-%. What might be expected if lignocellulose rather than pure cellulose is the feedstock?

6.5 The owner of a swine facility containing 10,000 pigs proposes to produce biogas from the manure. Assuming volatile matter is 80% of the dry solids, estimate the daily methane production rate, (m^3/day), and the annual energy production, (GJ/yr).

Processing of Biorenewable Resources into Chemicals and Fuels

7.1 Introduction

A variety of processes have been developed for converting biorenewable resources to chemicals and fuels including fermentation of sugars and starches, fractionation of lignocellulose to simple sugars, lipid extraction from oil seeds and terpene-rich plant materials, fast pyrolysis, direct liquefaction of biomass by thermal processes, and indirect liquefaction of biomass-derived syngas. The following sections describe each of these conversion processes, as well as secondary conversion processes to value-added products.

7.2 Fermentation of Sugars and Starches

Fermentation is a biological process in which enzymes produced by microorganisms catalyze energy-releasing reactions that break down complex organic substrates. Anaerobic conditions, characterized by the exclusion of oxygen, are most common, but aerobic processes, in which oxygen is present, are also possible. The French chemist Louis Pasteur established the field of microbiology with his efforts to help the brewery industry improve its product. He introduced the word fermentation, from the Latin *fermentare* (to boil), to describe microbial activity that releases gas as foam.

Fermentation can produce a wide variety of chemicals, although most of them are organic acids and alcohols. Most microorganisms used in commercial fermentations require six-carbon sugars (hexoses) or disaccharides as substrates, although the microbial world contains organisms that can breakdown virtually any organic compound. Recent technological advances have made possible the fermentation of five-carbon sugars (pentoses).

Several factors limit the use of fermentation technology in the production of chemicals. Production rates by microorganisms in aqueous media of low-solids volume are inherently low. The microorganisms are sensitive to both inhibitors and operating conditions, especially temperature and pH. Most fermentations require aseptic conditions, which can be difficult to achieve in large-scale operations. Recovery of water-soluble products from dilute solutions can be expensive. Effluent usually contains high biological oxygen demand (BOD) and requires wastewater treatment before discharge.

A major emphasis in the fermentation industry is the production of ethanol, which is marketed to both fuel and beverage industries. Table 7.1 lists ethanol fermentation yields for a variety of carbohydrate feedstocks. The maximum theoretical yield of ethanol is 0.51 (mass ethanol/mass carbohydrate, corresponding to 51% of the carbohydrate converted to ethanol on a mass basis), the balance of the yield being CO_2. Typically about 5–12 wt-% of the carbohydrate is converted to cells; thus, not more than 47 wt-% of the fermented carbohydrate is converted to ethanol. Of course other products can also be produced by fermentation, and these are described later in this chapter.

The following sections explore the technologies by which carbohydrates, including sugar, starch, and lignocellulose, are converted to co-products. Regardless of the carbohydrate, the following processes are involved: release of simple sugars from the carbohydrate, fermentation to organic acids or alcohols, distillation of the fermentation broth to recover the fermentation products, and utilization of co-products. Carbohydrates differ in how the simple sugars are released and in how co-products are utilized.

Table 7.1. Ethanol yields from various biorenewable resources

Feedstock	Yield (L/t)
Apples	64
Barley	330
Cellulose	259
Corn	355–370
Grapes	63
Jerusalem artichoke	83
Molasses	280–288
Oats	265
Potatoes	96
Rice (rough)	332
Rye	329
Sorghum (sweet)	44–86
Sugar beets	88
Sugarcane	160–187
Sweet potatoes	125–143
Wheat	355

Source: D. L. Klass, 1998, *Biomass for Renewable Energy, Fuels, and Chemicals*, Academic Press.

7.2.1 Sugar Crops

Traditional sugar crops used for fermentation include apples, grapes, other fruits, sugar cane, sugar beets, and sweet sorghum. Molasses, which is the residual syrup remaining from crystallization of sugar from sugar cane and sugar beets, is also a common feedstock. Even pulp and papermill sludges contain as much as 40–50 wt-% glucose. These sugars can be directly fermented by the yeast *Saccharomyces cerevisiae*, which contains enzymes that hydrolyze disaccharides to simple sugars and catalyze the fermentation of four hexoses: glucose, mannose, fructose, and galactose. Preparing these feedstocks for fermentation is relatively simple. For example, in the case of sugar cane, which contains 20–30 wt-% sugar, the cane stalks are chopped and crushed and washed with hot water to remove the sugar.

The fermentation broth is adjusted to less than 22 wt-% of sugar to avoid inhibition of yeast activity. Nutrients such as ammonium sulfate are added as needed to make up deficiencies in the original feedstock. The fermentation broth (or beer, as it is also called) exiting the process contains 6–10 vol-% ethanol.

Brazil has built a transportation sector that relies heavily upon ethanol fermented from sugar crops. However, in the United States, sugar subsidies make it too expensive for use in the production of ethanol.

7.2.2 Starch and Inulin Crops

Starch is a polymer that accumulates as granules in many kinds of plant cells where they serve as a storage carbohydrate. Mechanical grinding readily liberates starch granules. The hydrogen bonds between the basic units of maltose in this polymer are easily penetrated by water, making depolymerization and solubilization relatively easy.

Hydrolysis, the process by which water splits a larger reactant molecule into two smaller product molecules, is readily accomplished for starch. Acid-catalyzed hydrolysis in "starch cookers" at temperatures of 150–200°C proceeds to completion in seconds to minutes. In recent years enzymatic hydrolysis has supplanted acid hydrolysis due to higher selectivity.

Starch is a glucose polymer with two main components: *amylose*, a liner polymer of glucose with α–1,4 linkages, and *amylopectin*, a branched chain including α–1,6 linkages at the branch points. Thus, enzymatic hydrolysis of starch, known as *saccharification,* requires two enzymes. The enzyme amylase hydrolyzes starch to maltose in a process known as *liquefaction.* The enzyme maltase hydrolyzes maltose into glucose in a process known as saccharification. The consumption of either acid or enzymes for starch hydrolysis is less than 1:100 by weight, making the cost of hydrolysis only a small part of the cost of starch fermentation.

Inulin, like starch, is a storage carbohydrate, but its basic unit is fructose rather than glucose. It is commonly found in tuber crops such as dahlias and Jerusalem artichokes. It is also easily depolymerized and solubilized by both acid- and enzyme-catalyzed hydrolysis. Although potentially an important fermentation feedstock, reversion of fructose to undesired oligofructosans, and the lack of widespread cultivation of inulin crops, has limited its use.

Cereal grains, such as corn, wheat, and barley, are the most widely used sources of starch for fermentation. The cell walls of grains must be disrupted to expose starch polymers before they can be hydrolyzed to fermentable sugars (that is, *monosaccharides* and *disaccharides*). Grain starch consists of 10–20 wt-% amylose and 80–90 wt-% amylopectin, both of which yield glucose or maltose on hydrolysis. Although the amylose is water soluble, the amylopectin is insoluble and requires a "cooking" operation to solubilize it prior to hydrolysis.

Cereal grains also contain other components that may be of sufficient value to recover along with the starch, such as protein, oil, and fiber. For example, gluten, a mixture of plant proteins occurring in cereal grains (chiefly wheat and corn) is of value as an adhesive and animal feed. If these components are to be separately recovered, extensive pretreatment, known as wet milling, is required before the starch is hydrolyzed and fermented. Under some circumstances separation of plant components is not economically justified; simpler dry milling is employed to release starch polymers, and the whole grain is fermented. Of the 3.8 gigaliters (1.0 billion gal) of fuel ethanol produced annually in the United States, about one-third comes from dry milling plants while the remaining two-thirds is the result of wet milling.

The pretreatments of dry milling and wet milling are distinct in their operations, as detailed in the following paragraphs. The subsequent processes of hydrolysis, fermentation, and distillation, however, are very similar for the two kinds of corn milling operations. Either relatively pure starch from wet milling, or a starch-rich "mash" from dry milling, is treated with the enzyme amylase and heated to around 93°C to partially hydrolyze the starch and form a mixture of oligosaccharides and polysaccharides known a *dextrin*. The mash of dextrin is cooled to between 60 and 70°C, the pH adjusted to 3–5, and the enzyme glucoamylase added. This final hydrolysis step yields glucose.

The capital investment for dry milling is less than that for a comparably sized wet-milling plant. However, the higher value of its by-products, greater product flexibility, and simpler ethanol production can make a wet-milling plant a more profitable investment.

7.2.2.1 Dry Milling of Corn

Dry milling is essentially a simple grinding procedure. In the food industry a series of grinding and screening steps are employed. First, the corn is cleaned and then conditioned for 24 hours to increase moisture content to 15–20 wt-%. The kernels then pass through roller mills or crushers to separate the kernel into germ, endosperm, and fibrous hull by gravity. The germ contains 18–25 wt-% oil, which can be recovered by solvent extraction. The starchy particles of the endosperm can be further ground and sieved to three products distinguished by their particle size: flour, cornmeal, and grits. About 75 wt-% of the kernel becomes flour, cornmeal, or grits, 15 wt-% is recovered as oil, and 11 wt-% is used as fibrous feed. The starchy components can be used in food products or saccharified for fermentation.

For production of fuel ethanol by dry milling, illustrated in Fig. 7.1, separation into various products is not necessary. Kernels are ground in a roller mill to grain meal consistency to expose the starch. The meal is slurried with water to form a mash, which is then fermented to ethanol. The fibrous residue remaining upon completion of fermentation is recovered from the base of the beer stripping column, mixed with yeast and other unfermented residues, and dried to a co-product known as *distillers' dried grains and solubles (DDGS)*. This co-product, containing about 25 wt-% protein and residual oil, is a valuable feed for cattle. Profitability of a corn-to-ethanol plant is strongly tied to the successful marketing of DDGS.

A typical dry-milling plant will produce about 9.5–9.8 L (2.5–2.6 gal) of ethanol per bushel of corn processed. Yields of co-products per bushel of corn are

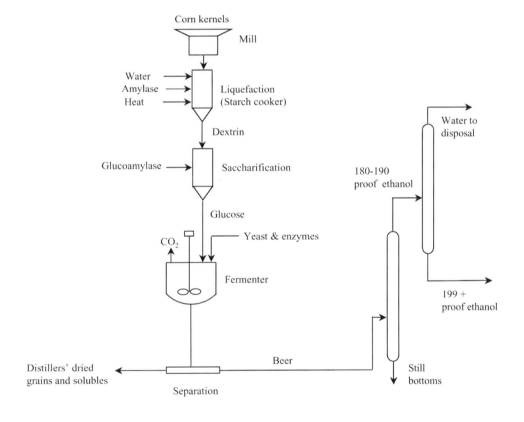

FIG. 7.1. Dry milling of corn.

7.7–8.2 kg (17–18 lb) of DDGS and 7.3–7.7 kg (16–17 lb) of CO_2 evolved from fermentation, the latter of which can be sold to the carbonated beverage industry. As a rule of thumb, the three products are produced in approximately equal weight per bushel with each accounting for approximately one-third of the initial weight of the corn.

7.2.2.2 Wet Milling of Corn

Wet milling has the advantage that it separates plant components into carbohydrate (starch), lipids (corn oil), a protein-rich material (gluten), and fiber (hulls). This gives a company access to higher-value markets as well as providing flexibility in the use of starch as a food product or in the production of fuel ethanol.

The wet-milling process is illustrated in Fig. 7.2. The corn is cleaned and then conveyed into steep tanks where it is soaked in a dilute solution of SO_2 for 24–36 hours, swelling and softening the corn kernels. Some of the protein and other compounds are dissolved in the resulting corn-steep liquor, which is an inexpensive source of nitrogen and vitamins.

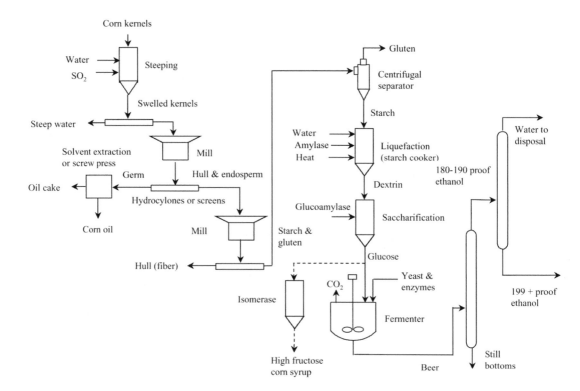

FIG. 7.2. Wet milling of corn.

After separating the corn from the steep liquor, the wet kernels are coarsely ground to release the hull and germ from the endosperm. Hydrocyclones or screens separate the germ from the rest of the components. After drying, oil is extracted from the germ by either solvents or a screw press, leaving a residual oil cake. The hull and endosperm pass through rotating disc mills. The mills grind the endosperm into fine fractions of starch and gluten and the hull into coarser fiber particles that can be screened out from the finer fractions. Centrifugal separators separate the lighter gluten from the starch.

The starch can be used directly as a food product or for industrial manufacturing processes, especially papermaking. The starch can also be converted to monosaccharides for the production of food or fuel, depending on relative market demand. Saccharification by amylase enzymes yields corn syrup, a glucose solution that can be directly fermented to fuel ethanol. Alternatively, treated with isomerase enzymes, the glucose is partially converted to fructose to yield a liquid sweetener known as *high fructose corn syrup (HFCS)*. In plants that can alternate between fuel ethanol and HFCS production, relatively more ethanol is produced in the winter, while relatively more HFCS is produced in the summer.

The gluten product, known as corn gluten meal, contains 60% protein and is used primarily as poultry feed. The fiber from the hulls is combined with other by-products, such as the oil cake, steep-water solubles, and excess yeast from stillage, and is dried and sold as corn gluten feed. Containing 21% or more of protein, it is primarily used as feed for dairy cattle.

A typical wet milling plant will produce 9.5–9.8 L (2.5–2.6 gal) of ethanol per bushel of corn processed. Yields of other co-products per bushel of corn are 0.7 kg (1.7 lb) of corn oil, 1.4 kg (3 lb) of corn gluten meal (60% protein), 5.9 kg (13 lb) of corn gluten feed (21% protein), and 7.7 kg (17 lb) of CO_2. Like dry milling, the three products of ethanol, feed, and carbon dioxide are produced in approximately equal weight per bushel, with each accounting for approximately one-third of the initial weight of the corn.

7.2.3 Fermentation to Products Other Than Ethanol

Only a small fraction of commercial production of organic compounds comes from microbial processes. With the exception of ethanol, most fermentation products are specialty chemicals including carboxylic acids, amino acids, antibiotics, industrial and food enzymes, and pharmaceuticals. Examples of organic chemicals produced from fermentation are listed in Table 7.2.

Virtually all commodity organic chemicals are currently produced from petroleum by virtue of its low price compared to biorenewable resources and the economics of scale that are achieved in modern petrochemical plants. Only a few fermentation products can be considered commodity chemicals (world-wide production exceeding

Table 7.2. Examples of chemical products from fermentation of carbohydrates

Chemical	Substrate(s)	Microorganism(s)
Acetic acid	Various sugars	*Acetobacter aceti*
		Clostridium
		Thermoaceticum
		Pachysolen tannophilus
Acetone	Various sugars	*Clostridium* sp.
2,3-Butanediol	Various sugars and acids	*Aerobacter aerogenes*
		Bacillus polymyxa
		Klebsiella oxytoca
		K. pneuminiae
n-Butanol	Various sugars and organics	*Clostridium* sp.
Butyraldehyde	Glucose	*Clostridium acetobutylicum*
Butyric acid	Various sugars	*Clostridium* sp.
Citric acid	Various sugars	*Aspergillus niger*
		Saccharomycopsis lipolytica
Ethanol	Various sugars	*Kluyberomyces* sp.
		Candida utilis
		Saccharomyces cerevisiae
		Zymomonas mobilis
C_{12}-C_{20} fatty acids	Sucrose	*Bacteria, mold, yeast*
Gluconic acid	Glucose	*Aspergillus niger*
		Gluconobacter suboxydans
Isopropyl alcohol	Various sugars	*Clostridium* sp.
Itaconic acid	Glucose, sucrose	*Aspergillus itaconicus*
		Ustilago zeae
		A. terreus
Linoleic acid	Glucose, lactose	*C. curvata*
Linolenic acid	Various sugars	*Mortierella ramammiana*
Oleic acid	Glucose, lactose	*C. curvata*
Palmitic acid	Glucose, lactose	*C. curvata*
Propanediol	Algal biomass, glucose	*Clostridium pasteurianum*
		C. thermosaccharolyticum
n-Propanol	Glucose	*Clostridium fallax*
Propionic acid	Various sugars	*Clostridium* sp.
		Propionibacterium shermanii
Pyruvic acid	Glucose	*Pseudomonas aeruginosa*
Sorbitol	Sucrose	*Zymomonas* sp.
Stearic acid	Glucose, lactose	*C. curvata*
Succinic acid	Various sugars	Many species

Source: S. A. Leeper, T. E. Ward and G. F. Andrews, 1991, "Production of Organic Chemicals via Bioconversion: A Review of the Potential" (Idaho Falls, ID: Report EGG-BG-9033, Idaho National Engineering Laboratory).

50,000 t/yr): ethanol, monosodium glutamate, citric acid, lysine, and gluconic acid. Ethanol leads this group (15 million ton/yr) by virtue of various tax incentives for its manufacture from renewable resources. New developments in fractionation of lignocellulose to simple sugars, and environmental considerations, may ultimately increase the production of commodity chemicals, such as lactic acid, 2,3 butanediol, itaconic acid, and propanediol, from biorenewable resources.

Only a relatively small number of organic compounds are produced as the primary products of anaerobic fermentation. These include a few simple alcohols and acids, a single ketone (acetone), and two gases of interest as fuel, methane and hydrogen. All of these products are derived from the single intermediate pyruvate. The primary substrates for most of these fermentations are monosaccharides or oligosaccharides, sugars consisting of two or three monosaccharides. Because transport of nutrients and metabolic products through the cell walls of microorganisms is limited to low-molecular weight compounds, the higher weight oligosaccharides and polysaccharides must be depolymerized outside the organisms to low molecular weight sugars before they can be metabolized. Microorganisms accomplish this by secreting enzymes such as amylase or cellulase to break down starch and cellulose, respectively, to sugars that can be transported into the cell.

The following paragraphs describe how various commodity chemicals could be produced through fermentation.

2,3-Butanediol

The heating value of butanediol (27,198 J/g) is close to that of ethanol (29,055 kJ/kg) and methanol (22,081 kJ/kg), making it a potentially attractive fuel. Many strains of microorganisms are able to ferment five- and six-carbon sugars to 2,3-butanediol including *Bacillus polymyxa* and *Klebsiella (Aerobacter) pneumoniae*. Under anaerobic conditions approximately equimolar amounts of ethanol and 2,3-butanediol are produced by all strains. Limited aeration decreases ethanol production, and 2,3-butanediol becomes the major or even sole product. Final product concentration for *B. polymyxa* is only 2–3% w/v, while *K. pneumoniae* can accumulate to 6–8% w/v. Recovery of 2,3-butanediol is difficult for several reasons, including high boiling point, high heat of vaporization, and its hydrophilic nature.

Aliphatic Acids

Aliphatic acids are derived from straight or branched-chain organic compounds that include alkenes, alkanes, or alkynes. Some short-chain aliphatic acids, such as acetic and propionic acids, can be produced as major products of fermentation. More typically, though, a mixture of acids and alcohols is produced. These are the same acetogenic bacteria as are involved in anaerobic digestion as described in Chapter 6, except the methanogenic step is not allowed to proceed. The accumulating organic acids are toxic to the microorganisms that create them. Hydroxides or carbonates of sodium, potassium, magnesium, calcium, or ammonium are added to the fermentation broth to neutralize organic acids as they are formed. For example, *Clostridium thermoaceticum* and *Clostridium thermoautotropicum* ferment fructose and a few six-carbon sugars to two moles of acetate by decarboxylation of pyruvate, while the third mole is produced by reduction and incorporation of CO_2.

High base consumption is an inherent feature of organic acid fermentation; the resulting salts of aliphatic acids (acetate) must be decomposed and the cations recovered if the process is to be cost effective. One approach is to pyrolyze the resulting acetate at 300°C to form ketones such as 2-propanone, 2-butanone, 2-pentanone, and 3-pentanone. The mixed ketones, which have a high energy content, a high octane rating, and a boiling range compatible with gasoline, have been proposed as transportation fuel. Alternatively, the ketones can be hydrogenated to mixed alcohols: 2-propanol, 2-butanol, 2-pentanol, and 3-pentanol.

Lactic Acid

Lactic acid, a 3-carbon molecule, is used in the production of *polylactide (PLA) resin*, a biodegradable polymer expected to compete with polyethylene and polystyrene in the synthetic fibers and plastics markets. Lactic acid is currently produced by milling corn, separating the starch, hydrolyzing the starch to glucose, and anaerobically fermenting the glucose to lactic acid with *Bacillus dextrolacticus* or *Lactobacillus delbrueckii*. Esterification with ethanol produces ethyl lactate, which can be polymerized to polylactate resin.

Succinic Acid

Succinic acid is used in producing food and pharmaceutical products, surfactants and detergents, biodegradable solvents and plastics, and ingredients to stimulate animal and plant growth. Although it is a common metabolite formed by plants, animals, and microorganisms, its current commercial production of 15,000 tons per year is from petroleum. However, the recently discovered rumen organism *Actinobacillus succinogenes* produces succinic acid with yields as high as 110 g/L, offering prospects for producing this chemical from biorenewable resources. In contrast to most other commercial fermentations, the process consumes CO_2 and integrated with a process like ethanol fermentation, succinic acid production could contribute to reductions in greenhouse-gas emissions.

Optimum yields occur under pH conditions where succinate salt rather than free acid is produced. Thus recovery entails concentration of the salt, conversion back to free acid, and polishing of the acid to the desired purity. Downstream purification can account for 60–70% of the product cost. One recovery approach uses calcium hydroxide to precipitate calcium succinate from the fermentation broth. Acidification of the precipitate yields gypsum. Disadvantages include handling and disposing of large amounts of wet gypsum. Another recovery approach employs simultaneous electrodialysis, acidification, and crystallization. Sodium succinate and other ionic species are transported across the ion-exchange membranes and separated from concentrated sugars, proteins, and amino acids.

7.3 Conversion of Lignocellulosic Feedstocks to Sugar

Much of the carbohydrate in plant materials is structural polysaccharides, providing shape and strength to the plant. The hydrolysis of polysaccharides in cell walls is more difficult than the hydrolysis of storage polysaccharides such as starch. This structural material, known as *lignocellulose*, is a composite of cellulose fibers embedded in a cross-linked lignin-hemicellulose matrix. Depolymerization to basic plant components is difficult because lignocellulose is resistant to both chemical and biological attack.

A variety of physical, chemical, and enzymatic processes have been developed to fractionate lignocellulose into the major plant components of hemicellulose, cellulose, and lignin. The hemicellulose fraction is readily hydrolyzed to pentoses (five carbon sugars), but pentoses are difficult to ferment. The cellulose exists in amorphous and crystalline forms, both of which hydrolyze to hexoses (six carbon sugars). Crystalline cellulose is recalcitrant to hydrolysis. However, the resulting hexoses are readily fermented. Distillation can recover the desired products of fermentation. Lignin, which is not susceptible to biological transformation, can be chemically upgraded or, more frequently, simply burned as boiler fuel. The steps of pretreatment, hydrolysis, fermentation, and distillation in the production of biobased products from lignocellulose are described in the following sections.

7.3.1 Pretreatment

Pretreatment is one of the most costly steps in the conversion of lignocellulose to sugars, accounting for about 33% of the total processing costs. Pretreatments often produce biological inhibitors, which impact the cost of fermenting the resulting sugars. Accordingly, much attention is directed at developing low-cost and effective pretreatments.

An important goal of all pretreatments is to increase the surface area of lignocellulosic material, making the polysaccharides more susceptible to hydrolysis. Thus, *comminution*, or size reduction, is an integral part of all pretreatments. Primary size reduction employs hammer mills to produce particles that can pass through 3 mm screen openings. The process has relatively modest energy requirements, ranging from 24,000 kJ per dry ton for wheat straw, to 200,000 kJ per dry ton for aspen wood. However, energy consumption increases exponentially with decreasing particle size. If the subsequent hydrolysis process requires further improvements in accessibility and susceptibility of polysaccharides, alternatives to finer milling are usually employed.

As described in the next section, three types of hydrolysis have been developed for releasing sugars from lignocellulose: two that employ mineral acids and one based on enzymes. The molecules of mineral acids, which are small compared to the pore volume of plant tissue, diffuse deeply into lignocellulosic material; thus primary size reduction produces sufficient surface area for acid hydrolysis. Additional pretreatment prior to acid hydrolysis, though, is often practiced to improve the yield of pentose sugars, as subsequently explained. Enzymatic hydrolysis always requires pretreatment beyond size reduction. Cellulase enzymes are large proteins with molecular weights ranging from 30,000 to 60,000 and are thought to be ellipsoidal with major and minor dimensions of 30 and 200 Å. Typically, only 20% of the pore volume of plant tissue is accessible to these large molecules. Without additional pretreatment, sugar yields from enzymatic hydrolysis are less than 20% of theoretical, whereas pretreatment can increase yields to 90% and higher.

The mechanisms by which pretreatments improve the digestibility of lignocellulose are not well understood. Pretreatment effectiveness has been correlated with removal of hemicellulose and lignin. Lignin solubilization is beneficial for subsequent hydrolysis but may also produce derivatives that inhibit enzyme activity. Some pretreatments are thought to reduce the crystallinity of cellulose, which improves reactivity, but this does not appear to be the key for many successful pretreatments. The large variety of pretreatment processes developed can be broadly classified as biological, alkaline, steam explosion, pre-hydrolysis, ammonia fiber explosion, and treatment with organic solvents.

Biological pretreatments employ microorganisms that produce lignin-degrading enzymes (ligninase). As lignin is decomposed, cellulose and hemicellulose are released from the lignocellulosic matrix. The exploitation of ligninase-producing microorganisms has been little developed and faces several hurdles. These include long reaction times (measured in weeks) and the fact that many ligninolytic microorganisms grow on the resulting sugars and produce cellulases and hemicellulases, thus degrading yields.

Alkali metal hydroxides not only break lignin-hemicellulose bonds but also dissolve lignin and hemicelluloses. At very high concentrations (5–20 wt-%) they also swell cellulose. The degradative and dissolving action of hot sodium hydroxide solutions is the basis of the kraft pulping process described in Chapter 8. However, alkali pulping creates long cellulose fibers, complicating subsequent biochemical processing, and destroying hemicelluloses, a significant source of fermentable sugars. The process can be adjusted to provide a pretreatment that is compatible with subsequent enzymatic hydrolysis of herbaceous biomass, such as wheat straw, corn stover, and some hardwoods. It has not been achieved for softwoods. A barrier to commercial use of this pretreatment is the significant amount of chemicals consumed in neutralizing the acidic carboxylic groups in biomass.

Steam explosion involves saturation of the pores of plant materials with steam followed by rapid decompression; the explosive expansion of steam reduces the plant material to separated fibers, presumably increasing the accessibility of polysaccharides to subsequent hydrolysis. Early research focused on the mechanical disruption of plant tissues and employed relatively high temperatures (220–270°C) and short residence times (40–90 s). Conditions and equipment for steam explosion are similar to the commercial masonite process developed in the 1930s for the production of fiberboard.

In sugar production from lignocellulose, recent research suggests that more important than mechanical disruption are the chemical changes that occur prior to the rapid decompression step. These include hydrolysis of hemicellulose and the condensation of lignin into small droplets. The removal of hemicelluloses and concurrent condensation of lignin creates numerous large pores in hardwoods and herbaceous biomass, which allow penetration of cellulase enzymes to the cellulose fibers. Researchers discovered that comparable hydrolysis of hemicellulose with reduced pyrolytic decomposition could be achieved at lower temperatures (190–210°C) by use of longer residence times (3–15 min). Hydrolysis of hemicellulose yields mostly pentoses. *Xylan,* the most common polysaccharide in hemicelluloses, consists of xylose units. Thus, the dominant pentose in the hydrolysate from hemicellulose is xylose followed by arabinose.

This steaming of biomass for several minutes is called *autohydrolysis* based on the hypothesis that acidic compounds released from the biomass are responsible for the process. Although this hypothesis has never been proven, the name persists in the literature. The main drawback of autohydrolysis is the relatively low yield of hemicellulosic sugars (about 50%) due to partial hydrolysis and pyrolytic decomposition at high temperatures. Furthermore, the process is not effective with softwoods.

Addition of small amounts of mineral acid, usually H_2SO_4, improves hydrolysis of hemicellulose at reduced temperatures. This process is variously known as *prehydrolysis, dilute acid pretreatment,* or *acid-catalyzed steam explosion.* The comminuted biomass is treated with 1 wt-% H_2SO_4 and incubated at 140°C for 30 minutes, or at 160°C for as little as 5–10 minutes, to achieve complete hemicellulose removal, which increases enzymatic digestibility of the remaining cellulose to as high as 90%.

Alternatively, SO_2 gas can be added in the amount of 2–3 wt-% to moist biomass chips and heated to 150°C for 20 minutes to hydrolyze hemicellulose. The sulfur dioxide rapidly diffuses into biomass pores before it is converted into H_2SO_4, providing superior performance compared to the direct use of an acid catalyst. It is far less corrosive than mineral acids. Steam pretreatment of hardwoods using SO_2 can recover more than 80% of the hemicellulose as monomers and more than 90% of the lignin by alkali washing, while enzymatic hydrolysis of the pretreated biomass can convert 100% of the cellulose to fermentable sugars. The same process can recover 65% of the hemicellulose as monomers, 80% of the lignin by a combination of both alkali- and peroxide-washing, and complete conversion of the cellulose by enzymatic hydrolysis. Corncobs, which have very high hemicellulose content, are particularly easy to digest. Pre-hydrolysis at 150°C converts the hemicellulose to xylose after only 5 minutes.

The resulting prehydrolysates contain soluble dextrins and sugars, acetic acid, furfural, low-molecular weight phenols derived from lignin, and other fermentation inhibitors. The acetic acid is inhibitory to fermentation above a threshold value of 0.5 g/L. Hardwoods such as aspen release as much as 6–10 g/L of acetic acid during pre-hydrolysis, while softwoods, which have fewer acetate groups in their hemicellulose, produce 2–4 g/L. Acetic acid and other volatile compounds can be readily removed from the pretreated biomass by steam to the required 0.5 g/L level or less. Furfural and other volatile compounds are also stripped out.

Pre-hydrolysis also yields furfural from xylose, which can be recovered during the flash cooling that occurs during pressure letdown at the discharge of the reactor. *Furfural*, which is a valuable solvent in oil refining and is also a base for furan resins, has been commercially produced from a variety of agricultural residues, including corncobs, cornstover, and oat hulls.

A disadvantage of pre-hydrolysis is the need to neutralize the acidified biomass. Although calcium hydroxide (lime) is a relatively inexpensive base, the resulting $CaSO_4$ (gypsum) is of low value and represents a waste disposal problem. Also, some sugar decomposition invariably occurs.

Ammonia fiber explosion (AFEX) is similar to steam explosion except that liquid ammonia is employed. Pressures exceeding 12 atm are required for operation at ambient temperature, with ammonia loadings in the range of 1.0–2.5 kg per kg dry biomass. The mixture is incubated for several minutes to up to an hour to enable ammonia to penetrate the lignocellulosic matrix. Hydrolysis yields from AFEX-treated agricultural residues are 80–90% of theoretical, superior to that achieved by pre-hydrolysis. Furthermore, no gypsum by-product is generated. The process has not been successfully applied to either hardwoods or softwoods.

Organic solvents have been used as pretreatments to remove lignin from biomass. Lignin can adsorb significant amounts of cellulase enzymes employed for enzymatic hydrolysis of cellulose. Thus, lignin removal can decrease the requirement for costly enzymes in downstream processing. Selective removal of lignin can be accomplished by addition of organic solvents (usually lower alcohols) to the acidic or alkaline aqueous solutions of pulping or pretreatment processes.

7.3.2 Hydrolysis

Three basic methods for hydrolyzing structural polysaccharides in plant cell walls to fermentable sugars are available: concentrated-acid hydrolysis, dilute-acid hydrolysis, and enzymatic hydrolysis. The two acid processes hydrolyze both hemicellulose and cellulose with very little pretreatment beyond comminution of the lignocellulosic material to particles of about 1 mm size. The enzymatic process must be preceded by extensive pretreatment to separate the cellulose, hemicellulose, and lignin fractions.

Concentrated-acid hydrolysis is based on the discovery, over a century ago, that carbohydrates in wood will dissolve in 72% H_2SO_4 at room temperature leaving behind the lignin fraction. For fermentation, the solution of oligosaccharides is diluted to 4% H_2SO_4, and heated at the boiling point for 4 hours, or in an autoclave at 120°C for 1 hour to yield monosaccharides. Following neutralization with limestone, the sugar solution can be fermented. Concentrated acid hydrolysis is relatively simple and is attractive for its high sugar yields, which approach 100% of theoretical hexose yields.

Both H_2SO_4 and HCl have been considered for commercial development of concentrated acid hydrolysis. The low price of H_2SO_4 makes it attractive as a hydrolyzing agent. Nevertheless, the large volume of acid required, about equal to the weight of sugars produced, mandates its recovery and reuse. The recovery of H_2SO_4 is complicated by its high boiling point. Electrodialysis and solvent extraction are possible recovery options. Hydrochloric acid is significantly more expensive and corrosive than H_2SO_4 but its higher volatility presents opportunities for recovery by distillation. Evaporation under vacuum yields a high boiling azeotrope (18% HCl, 120°C). An *azeotrope* is a liquid mixture of two or more substances that retains the same composition in the liquid and vapor states, thus presenting difficulties in recovery of one component by distillation. Neutralization of the acid with limestone also generates a large waste stream of gypsum ($CaSO_4$). An ethanol plant would generate over 2 kg of gypsum per liter of ethanol produced, or 40,000 tons of wet gypsum for a 20 million L/yr plant.

Dilute-acid hydrolysis (about 1% acid by weight) greatly reduces the amount of acid required to hydrolyze lignocellulose. The process is accelerated by operation at elevated temperatures: 100–160°C for hemicellulose and 180–220°C for cellulose. Unfortunately, the high temperatures cause oligosaccharides released from the lignocellulose to decompose, greatly reducing yields of simple sugars to only 55–60% of the theoretical yield. The decomposition products include a large number of microbial toxins such as acetic acid and furfural, which inhibit fermentation of the sugars. The need for corrosion-resistant equipment and the low concentrations of sugars from some reactor systems also adversely impact the cost of sugars.

Early efforts in both concentrated- and dilute-acid hydrolysis of lignocellulose focused on recovery of hexoses, which are easily fermentable. Pentoses released from hemicellulose were considered of little value, and no attempt was made to recover them or prevent their degradation. Since biomass is a relatively expensive feedstock, recovery of both hexoses and pentoses is now considered essential to the economic viability of any lignocellulose-to-sugars process. Towards this end, acid-hydrolysis processes have incorporated pretreatments, not for the purpose of increasing cellulose accessibility but to separately recover pentose from hemicellulose to prevent its degradation under the harsh conditions of acid hydrolysis.

An example of a lignocellulose-to-ethanol plant based on concentrated acid hydrolysis with pretreatment to recover hemicellulose is illustrated in Fig. 7.3. Milled biomass is conveyed to a pre-hydrolysis (hemicellulose) reactor for the purpose of converting hemicellulose into pentose. The biomass is treated with a stream of 9% H_2SO_4 and dissolved glucose is recycled from the concentrated acid (cellulose) reactor. The slurry is heated to 100°C for 2–3 hours, which is sufficient to hydrolyze the hemicellulose. The liquid, containing both pentose and hexose, is pressed from the biomass, neutralized with limestone, filtered to remove the resulting gypsum, and sent to a fermentor. The pre-hydrolyzed biomass enters one or more stages of a concentrated-acid (cellulose) reactor where it is treated with 70% H_2SO_4 at 100°C for 2–4 hours. Hydrolysis yields hexose and lignin, which is

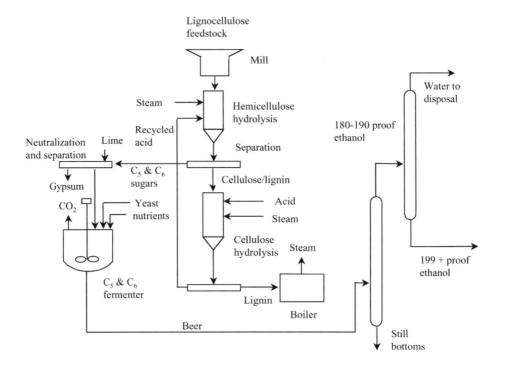

FIG. 7.3. Concentrated acid hydrolysis of lignocellulosic biomass.

removed by a filter press. The acidic sugar solution is recycled to the pre-hydrolysis reactor. When corn stover is used as feedstock, alcohol recovery is 315 L/ton of biomass.

A dilute-acid hydrolysis plant with pretreatment to recover hemicellulose is illustrated in Fig. 7.4. The process resembles the concentrated-acid plant except that pentoses are recovered and fermented separately from the hexose, and cellulose is hydrolyzed with dilute H_2SO_4. Biomass is impregnated with 1% H_2SO_4 and heated with saturated steam at 1.24 MPa for 4 min to pre-hydrolyze hemicellulose. The pentoses are separated from the residual lignocellulose by filtering, neutralized with limestone, and sent to a pentose fermentor. The residual lignocellulose is impregnated with 3% H_2SO_4 and heated with steam at 2 MPa for 4 minutes to hydrolyze the cellulose. The hexose solution is filtered from the lignin residue, neutralized with limestone, and fermented in a hexose fermentor. The lignin is used as boiler fuel. The yield of alcohol from poplar wood is 250 L/ton.

Enzymatic hydrolysis was developed to better utilize both cellulose and hemicellulose from lignocellulosic materials. Pretreatment solubilizes hemicellulose under milder conditions than those required for acid hydrolysis of cellulose. Subsequent enzymatic hydrolysis of the cellulose does not degrade pentoses released during pre-hydrolysis.

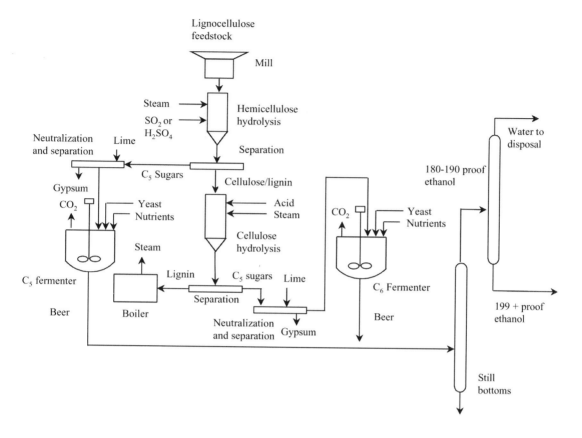

FIG. 7.4. Dilute acid hydrolysis of lignocellulosic biomass.

Cellulose is a homopolysaccharide of glucose linked by β-1,4'-glycosidic bonds. Thus, enzymatic hydrolysis of cellulose proceeds in several steps to break glycosidic bonds by the action of a system of enzymes known as cellulase. Native cellulose is hydrolyzed by the isoenzymes *cellobiohydrolase I* and *cellobiohydrolase II* to yield cellodextrins and cellobiose. The cellodextrins are further hydrolyzed to cellobiose, a disaccharide of glucose, by the isoenzymes *endoglucanase I* and *endoglucanase II*. The cellobiose is hydrolyzed to monosaccharides by *β-glucosidase*. The system of enzymes also usually contains hemicellulase to hydrolyze any hemicellulose not solubilized by pre-hydrolysis.

A variety of fungi and bacteria produce cellulases both aerobically and anaerobically. The aerobic mesophilic fungus *Trichoderma reesei* and its mutants have been the most intensely studied sources of cellulases. Other fungal cellulase producers include *T. viride*, *T. lignorum*, *T. koningii*, *Penicillium* spp., *Fusarium* spp., *Aspergillus* spp., *Chrysosporium pannorum*, and *Sclerotium rolfsii*. The problems with enzymatic hydrolysis include relatively low specific activity, low rates of conversion, and sensitivity to end-product inhibition. The low specific activity leads to high enzyme loading requirements: approximately 1 kg of enzyme is needed for hydrolysis of 50 kg of cellulose fibers. Conversion rates are as low as 20% in 24 hours; thus, up to seven days are required to digest lignocellulose. A technique described in the next section, the simultaneous hydrolyzing of cellulose and fermenting of hexose as it is released, can substantially overcome end-product inhibition.

7.3.3 Fermentation

The first step in a successful fermentation is removal of toxic compounds from the hydrolysate that would otherwise inhibit the growth of fermentation organisms. These toxic compounds include furfural and acetic acid, which are breakdown products from hydrolysis of hemicellulose. Traditional detoxification methods, such as the addition of activated carbon, extraction with organic solvents, ion exchange, ion exclusion, molecular sieves, over-liming, and steam stripping, can be costly. Another method under development is adaptation of the fermentation organisms to the inhibitory substances.

Numerous yeast species, including common baker's yeast, *Saccharomyces cerevisiae*, and two species of bacteria in the genus *Zymomonas*, efficiently ferment six-carbon sugars to ethanol and CO_2. These microorganisms are suitable for fermenting traditional sugar and starch crops, which yield hexoses upon processing. However, they are not able to ferment the pentoses released from lignocellulosic biomass upon hydrolysis of the hemicellulosic fraction. Efficient conversion of lignocellulosic biomass requires fermentation of pentoses, especially xylose and arabinose.

A variety of microorganisms can directly ferment pentose. Among wild-type yeast genera *Pichia stipitis*, *Candida shehatae*, and *Pachysolen tannophilus* are able to ferment both five- and six-carbon sugars. Maximum yields are on the order of 50 g/L compared to 150 g/L for hexose-fermenting *S. cerevisiae*. A few filamentous fungi, notably within the genera *Fusarium*, *Rhizopus*, and *Paecilomyces*, ferment both five- and six-carbon sugars to ethanol and CO_2, but at low rates and final concentrations of ethanol. Thermophilic bacteria such as *Clostridium thermohydrosulfuricum*, *C. thermosaccharolyticum*, and *C. thermocellum* are able to produce ethanol from both hexoses and pentoses, but they have low tolerances for end products. For example, the maximum yield of ethanol by *C. thermocellum* is less than 30 g/L.

Research into using wild-type bacteria and fungi to convert xylose waned during the 1980s. This was due in part to the fact that the disadvantages of low conversion rate and/or poor yield of these types of microorganisms were widely recognized, coupled with the fact that moderately productive xylose-fermenting yeasts had been identified. Recombinant DNA techniques are being employed to produce new strains of microorganisms with the desired trait of fermenting both hexoses and pentoses. For example, the bacteria *Escherichia coli* is able to convert both hexoses and pentoses to pyruvate, but the end product is acetic acid rather than ethanol. The bacteria *Z. mobilis* is able to convert pyruvate to ethanol but cannot ferment pentose. Using recombinant techniques, researchers transferred pyruvate decarboxylase and alcohol dehydrogenase from *Z. mobilis* to *E. coli* resulting in an organism able to ferment up to 90% of the sugars derived from lignocellulose, with final ethanol concentrations ranging from 40 to 58 g/L.

Three approaches have been developed for fermenting sugars released from lignocellulose: *separate hydrolysis and fermentation (SHF)*, *simultaneous saccharification and fermentation (SSF)*, and *direct microbial conversion (DMC)*.

The SHF process employs separate steps for the unit operations of pre-hydrolysis, enzymatic hydrolysis, and fermentation. The primary advantage of this approach is that by separating each step, undesirable interactions are avoided. Use of separate reactors for hydrolysis and fermentation allows these two processes to proceed at their optimal temperatures: 50°C for enzymatic hydrolysis and 20–32°C for fermentation. The disadvantage of this method is inhibition of the enzyme β-glucosidase by the hydrolysis product glucose, thus requiring lower solids loadings to obtain reasonable yields. Low sugar concentrations result in lower ethanol concentrations, which increase the cost of fermentation and subsequent product recovery.

The SSF process combines hydrolysis (saccharification) and fermentation to overcome the end-product inhibition that occurs during hydrolysis of cellobiose. By combining hydrolysis and fermentation in the same reactor, glucose is rapidly removed before it can inhibit further hydrolysis. The SSF process is illustrated in Fig. 7.5. The biomass feedstock is milled and then pre-hydrolyzed to yield a mixture of pentoses, primarily xylose and arabinose, and fiber. The mixture is neutralized with limestone and mixed with cellulase and hemicellulase enzymes (which are either purchased commercially or produced on site), yeast, and nutrients. The cellulose and any remaining hemicellulose are solubilized to hexose (glucose) and pentoses (xylose and arabinose) that are in turn immediately fermented to ethanol. The rate-limiting step is the hydrolysis of cellulose to glucose. The optimum temperature for the hydrolysis/fermentation reactor is a compromise between the optimum temperature for cellulase activity, and the optimum temperature for the yeast. Lignin is separated from the mixture and used as boiler fuel. The beer is distilled to ethanol in a process identical to that employed after sugar or starch fermentations.

Direct microbial conversion (DMC) combines cellulase production, cellulose hydrolysis, and glucose fermentation into a single step. The process is attractive in that it reduces the number of reactors, simplifies operation, and reduces the cost of chemicals. An example of commercially successful application of DCM is anaerobic digestion of sewage sludge and agricultural wastes into methane and carbon dioxide. To date, though, product yields are low, undesired metabolic by-products are produced, and product inhibition is common. Further development should improve the attractiveness of this approach.

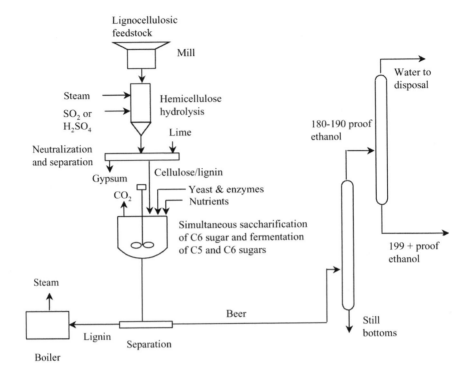

FIG. 7.5. Enzymatic hydrolysis of lignocellulosic biomass.

7.3.4 Distillation

Fermentation can produce gas (such as methane from anaerobic digestion), precipitate (such as calcium acetate during the production of aliphatic acids), or water-soluble compounds (such as ethanol). Gaseous or precipitated products are attractive because of the relative ease of separating them from the spent fermentation broth (beer). Distillation is an energy-intensive process used to recover water-soluble products of fermentation.

The first distillation yields 55% (v/v) ethanol and stillage bottoms, the latter of which contain significant protein in the case of whole grain fermentation. These stillage bottoms are marketed as animal feed under the name of *distillers' dry grains and solubles (DDGS)*. The second distillation produces an ethanol and water azeotrope containing 95–96% ethanol (190–192 proof). If essentially water-free ethanol is desired, purification beyond the azeotrope can be achieved by one of several methods: further distillation in the presence of a separating agent known as an *entrainer* (e.g., benzene, cyclohexane, heptane) that is subsequently recovered; absorption using corn grits or some other solid material; or a membrane-based operation.

Ethanol production has been criticized for consuming more energy than it produces. This is due, in part, to the energy consumption in the distillation process. Although there is basis for this criticism in older plants, modern plants pay close attention to energy consumption. Some plants are reported to use as little as 5.6 MJ of steam per liter of ethanol produced, with a total energy consumption of 11.1 to 12.5 MJ/L of product ethanol.

7.4 Lipid Extractives

Historically, *lipids* were defined as compounds of biological origin that are soluble in organic solvents. Lipids include terpenes and esters of fatty acids. *Terpenes* include a large number of hydrocarbons that occur naturally in many plants, some of which have been significant sources of biobased products for thousands of years. *Esters of fatty acids* produced in plants and a variety of microorganisms include waxes and triglycerides. The energy-rich *triglycerides,* esters of glycerol and long-chain fatty acids, are of particular interest in production of biobased products.

7.4.1 Triglycerides

Triglycerides, also known as *fats* and *oils*, are esters of glycerol and fatty acids. Recall that fatty acids are long-chain carboxylic acids containing even numbers of carbon atoms. The acid fractions of triglycerides can vary in chain length and degree of saturation. Fats, which are solid or semi-solid at room temperature, have a high percentage of saturated acids, whereas oils, which are liquid at room temperature, have a high percentage of unsaturated acids. Plant-derived triglycerides are typically oils containing unsaturated fatty acids, including oleic, linoleic, and linolenic acids.

A wide variety of plant species produce triglycerides in commercially significant quantities, most of it in their seeds. Table 7.3 is a partial listing of oilseed plants, their seed yields, and their oil yields. Average oil yields range from 150 L/ha for cottonseed to 814 L/ha for peanut oil, although intensive cultivation might double these numbers. Soybeans are responsible for more than 50% of world production of oilseed, representing 48–82 million barrels per year (bbl/year). However, the Chinese tallow tree, cultivated in the southern United States, has the potential for several-fold higher productivity than soybeans and is particularly attractive for its ability to grow on saline soils that are not currently used for agriculture.

Extraction of seed oil is relatively straightforward. The seeds are crushed to release the oil from the seed. Mechanical pressing is used to extract oil from seeds with oil content exceeding 20%. Solvent extraction is required for seeds of lower oil content. The residual seed material, known as meal, is used in animal feed.

Table 7.3. Seed yields and oil yields for selected oil seeds

Common Name	Species	Seed Oil Yield Average (L/ha)	Potential (L/ha)
Castorbean	*Ricinus communis*	449	1590
Chinese tallow tree	*Sapium sebiferum*	6270	6270
Peanut	*Arachis hypogaea*	814	1780
Safflower		599	940
Soybean	*Glycine max*	383	650
Sunflower	*Helianthus annuus*	571	1030

Source: E. S. Lipinsky, et al, 1984, "Fuels and Chemicals from Oilseeds," AAAS Selected Symposium 91, E. B. Shultz, Jr. and R. P. Morgan, eds. (Boulder, CO: Westview Press, Inc.).

Triglycerides are also recovered as a co-product of the pulping of pinewood by the kraft process, which is described in detail in Chapter 8. The esters of both fatty acids and resin acids are saponified to sodium salts and recovered as soap foam on the surface of the black (pulping) liquor. These salts are acidified to form a mixture of 30% fatty acids, 35% resin acids, and 35% unsaponifiable esters known as *tall oil*.

Microorganisms, including yeasts, fungi, and algae are also potential sources of triglycerides. Anaerobic yeasts and fungi accumulate triglycerides during the latter stages of growth when nutrients other than carbon begin to be exhausted. Nutrient deprivation known to trigger lipid accumulation includes nitrogen, phosphate, sulfur, and iron. However, conversion of sugar to lipid is generally only 15–24%, which would limit the process to very low-value sugar streams. Algae, which grow over a wide range of temperatures in high-salinity water, can produce as much as 60% of their body weight as lipids when deprived of key nutrients such as silicon (for diatoms) or nitrogen (for green algae). They employ relatively low substrate concentrations: on the order of 10–40 g/L. Product recovery is relatively simple (unlike in ethanol production) because of the sequestration of the oil in the algae. One proposed system is to build algae ponds in the desert Southwest where inexpensive flat land, water from alkaline aquifers, and CO_2 from power plants could be used to generate triglyceride-based fuel.

The higher viscosity and lower volatility of triglycerides compared to diesel fuel leads to coking of the injectors and rings of diesel engines. Chemical modification of triglycerides to methyl or ethyl esters yields excellent diesel-engine fuel. Biodiesel is the generic name given to these modified vegetable oils and animal fats. Suitable feedstocks include soybeans, sunflowers, cottonseeds, corn, groundnuts (peanuts), safflower seeds, rapeseed, waste cooking oils, and animal fats. Waste oils or tallow (white or yellow grease) can also be converted to biodiesel.

Transesterification describes the process by which triglycerides react with methanol or ethanol to produce methyl esters and ethyl esters, respectively, along with the co-product glycerol. For example, one triglyceride molecule reacts with three methanol molecules to produce one molecule of 1,2,3-propanetriol (glycerol) and three ester molecules:

triglyceride   methanol   glycerol   methyl ester

$$
\begin{array}{c}
\text{R1}-\overset{\overset{\displaystyle O}{\|}}{\text{C}}-\text{O}-\text{CH}_2 \\[2ex]
\text{R2}-\overset{\overset{\displaystyle O}{\|}}{\text{C}}-\text{O}-\text{CH} \;+\; 3\,\text{CH}_3\text{OH} \;\longrightarrow\; \text{HO}-\text{CH} \;+\; \text{R1}-\overset{\overset{\displaystyle O}{\|}}{\text{C}}-\text{O}-\text{CH}_3 \\[2ex]
\text{R3}-\overset{\overset{\displaystyle O}{\|}}{\text{C}}-\text{O}-\text{CH}_2
\end{array}
\tag{7.1}
$$

$$
+\; \text{R2}-\overset{\overset{\displaystyle O}{\|}}{\text{C}}-\text{O}-\text{CH}_3 \;+\; \text{R3}-\overset{\overset{\displaystyle O}{\|}}{\text{C}}-\text{O}-\text{CH}_3
$$

Near-quantitative yields of methyl (or ethyl) esters can be produced in one hour at room temperature using 6:1 molar ratios of alcohol and oil when catalyzed by 1% lye (NaOH or KOH).

The lye also serves as a reactant in the conversion of esters into salts of fatty acids. These salts are familiarly known as soaps and the process is called *saponification* (soap-forming). Small amounts of soap are also produced by the reaction of lye with fatty acids. Upon completion, the glycerol and soap are removed in a phase separator. A flow sheet for a biodiesel production facility is given in Fig. 7.6.

7.4.2 Terpenes

Terpenes are any of various isomeric hydrocarbons $(C_5H_8)_n$ where n is 2 or more. They were originally derived from turpentine, an oleoresin obtained from coniferous trees. Terpenes, abundant in a variety of plants, are responsible for the odor of pine trees and the color of carrots and tomatoes.

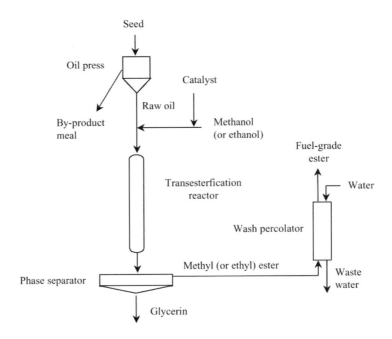

FIG. 7.6. Conversion of triglycerides to methyl (or ethyl) esters and glycerol.

Structurally, terpenes consist of two or more isoprene units joined together, which is called a *conjugated diene*, a compound containing two double bonds of carbon joined by one single bond. These compounds may have different degrees of unsaturation and a variety of functional groups. Terpenes range from relatively simple hydrocarbons to large polymeric molecules (polyisoprenes). Terpenes from conifers were the source of naval stores, the resinous substances used to maintain wooden ships and ropes. Polyisoprenes from the hevea rubber tree (*Hevea braziliensis*) or the guayule shrub (*Parthenium argentatum*) are the basis for natural rubber.

The hydrocarbon content of plants is a small fraction of the total plant material. Typical terpene content of mesophytic biomass species is only 0.5–1.5 wt-%. Thus, economic recovery requires tapping of living biomass, as is done with hevea rubber trees, or recovery as a co-product in the processing of other biomass components, such as in the production of turpentine as part of wood pulping.

7.5 Thermochemical Conversion

7.5.1 Fast Pyrolysis

Fast pyrolysis is the rapid thermal decomposition of organic compounds in the absence of oxygen to produce liquids, gases, and char. The distribution of products depends on the biomass composition and rate and duration of heating. Liquid yields as high as 78% are possible for relatively short residence times (0.5–2 s), moderate temperatures (400–600°C), and rapid quenching at the end of the process. Rapid quenching is essential if high-molecular weight liquids are to be condensed rather than further decomposed to low-molecular weight gases. Typical product yields for two kinds of wood are given in Table 7.4.

Pyrolysis liquid from fast pyrolysis is a low viscosity, dark-brown fluid with up to 15–20% water, as opposed to the black, tarry liquid resulting from slow pyrolysis or gasification. As indicated by Table 7.4, fast-pyrolysis liquid is a mixture of many compounds, although most can be classified as acids, aldehydes, sugars, and furans, derived from the carbohydrate fraction; and phenolic compounds, aromatic acids, and aldehydes, derived from the lignin fraction. The liquid is highly oxygenated, approximating the elemental composition of the feedstock, and thus highly unstable.

The liquid, despite its high water content, shows no appreciable phase separation. However, if an equal volume of water is added to the liquid, the high-molecular weight, largely aromatic compounds are precipitated. Since most of the aromatic compounds can be traced to the lignin content of the biomass, this precipitate is widely known as *pyrolytic lignin*. The pyrolysis liquids are highly corrosive due to the presence of organic acids derived primarily from the hemicellulosic content of the feedstock. The liquids also contain fine-particulate char. The higher heating values of pyrolysis liquids range between 17 MJ/kg and 20 MJ/kg, with liquid densities of about 1280 kg/m^3. Assuming conversion of 72% of the biomass feedstock to liquid on a weight basis, yield of pyrolysis oil is about 560 L/t.

The mechanism by which cellulose, hemicellulose, and lignin in biomass are converted into liquids is not fully understood. Rapid pyrolysis of pure cellulose yields *levoglucosan*, an anhydrosugar with the same molecular formula as the monomeric building block of cellulose: $C_6H_{10}O_5$. Addition of a small amount of alkali inhibits the formation of levoglucosan and promotes the formation of hydroxyacetaldehyde (glycolaldehyde). Pyrolysis of pure cellulose under slower heating rates and lower temperatures favors the formation of char rather than liquids. These observations suggest the multiple reaction pathways for pyrolysis of cellulose illustrated in Fig. 7.7. At low temperatures and slow heating rates, dehydration reactions are favored that ultimately convert the cellulose to char and water. At higher temperatures, depolymerization dominates, yielding levoglucosan as the primary product. The presence of alkali, however, catalyzes the dehydration route but yields hydroxyacetaldehyde instead of char and water if reaction products are removed fast enough. Similarly, hemicelluloses form furanoses and furans as primary reaction products while lignin forms monocyclic aromatics and non-condensed bicyclic aromatic materials with high phenolic content.

The mechanism by which reaction products are transported out of the reaction zone and recovered as liquids is also uncertain. Many of the reaction products, including levoglucosan, have very low vapor pressures, making vapor transport

Table 7.4 Analysis of products of fast pyrolysis

	White Spruce	Poplar
Moisture content, wt%	7.0	3.3
Particle size, μm (max)	1000	590
Temperature	500	497
Apparent residence time	0.65	0.48
Product yields, wt %, m.f.		
Water	11.6	12.2
Char	12.2	7.7
Gas	7.8	10.8
Pyrolytic liquid	66.5	65.7
Gas composition, wt%, m.f.		
H_2	0.02	—
CO	3.82	5.34
CO_2	3.37	4.78
CH_4	0.38	0.41
C2 hydrocarbons	0.20	0.19
C3+ hydrocarbons	0.04	0.09
Pyrolytic liquid composition, wt %, m.f.		
Organic liquid	66.5	65.7
Saccharides		
Oligosaccharides	—	0.70
Glucose	0.99	0.41
Other monosaccharides	2.27	1.32
Anhydrosugars		
Levoglucosan	3.96	3.04
1,6 anhydroglucofuranose	—	2.43
Cellobiosan	2.49	1.30
Aldehydes		
Glyoxal	2.47	2.18
Methylglyoxal	—	0.65
Formaldehyde	—	1.16
Acetaldehyde	—	0.02
Hydroxyacetaldehyde	7.67	10.03
Furans		
Furfural	0.30	—
Methylfurfural	0.05	—
Ketones		
Acetol	1.24	1.40
Alcohols		
Methanol	1.11	0.12
Ethylene glycol	0.89	1.05
Carboxylic acids		
Acetic acid	3.86	5.43
Formic acid	7.15	3.09
Water-soluble – total above	34.5	34.3
Pyrolytic lignin	20.6	16.2
Amount not accounted for (losses, water soluble phenols, furans, etc.)	11.4	15.2

Source: J. Piskorz, D. S. Scott and D. Radlein, 1988, *Pyrolysis Oils from Biomass*, E. J. Soltes and T. A. Milne, eds., ACS Symposium Series 376 (Washington, DC: American Chemical Society), 167–178.

Note: m.f. = moisture free.

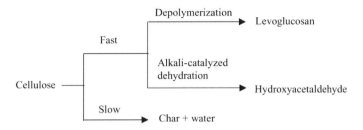

FIG. 7.7. Fast-pyrolysis reaction pathways.

problematic, but the feasibility of this approach has not been definitively disproved. Alternatives include transport of lower molecular weight compounds that condense to higher molecular weight compounds outside the reactor and the elutriation of fine liquid droplets (aerosols) from the reactor.

Pyrolysis liquid can be used directly as a substitute for heating oil. In some circumstances it is also suitable as fuel for combustion turbines or modified diesel engines. Recovery of high-value chemicals is another possibility, suggesting an integrated approach to production of both chemicals and fuel. For example, levoglucosan is obtained at high yields upon fast pyrolysis of pure cellulose or starch. Even woody or herbaceous biomass can yield significant levoglucosan if metal ions, particularly potassium, are washed from the material before it is pyrolyzed. Levoglucosan is considered a potential building block for synthesis of dextrin-like polymers, pharmaceuticals, pesticides, and surfactants. Microorganisms have been identified that can ferment levoglucosan to citric acid and itaconic acid, which may be an attractive source of these chemicals if the levoglucosan is obtained from inexpensive lignocellulosic materials.

Production of pyrolysis liquids and co-products is illustrated in Fig. 7.8. Lignocellulosic feedstocks, such as wood or agricultural residues, are milled to fine particles of less than 1 mm diameter to promote rapid reaction. The particles are entrained in an inert gas stream that transports the material to the pyrolysis reactor, shown here as a fluidized bed, which provides the prerequisite high heat-transfer rates. Within the reactor, the particles are heated rapidly and converted into condensable vapors, non-condensable gases, and solid charcoal. These products are transported out of the reactor into a cyclone operating above the condensation point of pyrolysis vapors where the charcoal is removed. Vapors and gases are transported to a direct-contact quench vessel where a spray of pyrolysis liquid cools and condenses the vapors. The non-condensable gases, which include flammable CO_2, H_2, and CH_4, are burned in air to provide heat for the pyrolysis reactor. A number of schemes have been developed for indirectly heating the reactor, including transport of solids into fluidized beds, or cyclonic configurations to bring the particles into contact with hot surfaces.

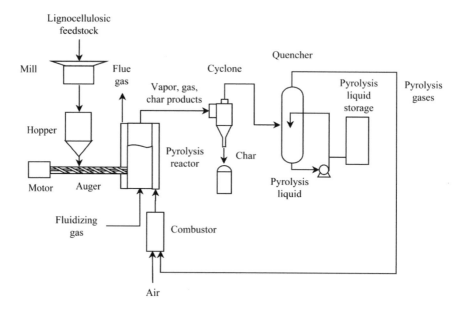

FIG. 7.8.   Fast-pyrolysis plant.

There are several problems with pyrolysis liquids. Phase-separation and polymerization of the liquids and corrosion of containers make storage of these liquids difficult. The high O_2 and water content of pyrolysis liquids makes them incompatible with conventional hydrocarbon fuels. Furthermore, pyrolysis liquids are of much lower quality than even Bunker C heavy fuel oil. Thus, upgrading pyrolysis liquids to more conventional hydrocarbon fuels, such as gasoline, is desirable.

Either traditional hydrotreatment, or emerging zeolite technology, can accomplish upgrading. *Hydrotreating*, commonly employed in the petrochemical industry, involves addition of H_2 at high pressure in the presence of a catalyst. Hydrotreating has the advantage of yielding high-quality products at maximum liquid yield. Disadvantages of hydrotreating include high hydrogen consumption and the need to operate at elevated pressures (13–17 MPa), both of which adversely affect the economics of the process. Upgrading of these liquids can also be accomplished at atmospheric pressure without reducing gases, by dehydration and decarboxylation over acidic zeolites. A yield of 17% of C_5-C_{10} hydrocarbons can be achieved from pyrolysis liquid derived from poplar wood.

Use as diesel fuel is complicated by long ignition delays and varying impacts of water, char, and volatiles content on relative burn rates. The viscosity, density, and surface tension of the pyrolysis liquids are significantly greater than those of diesel fuel, suggesting that poor atomization might hamper efficient ignition and combustion.

7.5.2 Direct Liquefaction

Direct conversion of biomass to liquids by the application of heat and pressure to aqueous mixtures was demonstrated over 100 years ago. The conditions for direct liquefaction are distinct from those for fast pyrolysis in several respects. Direct liquefaction occurs in an aqueous phase at temperatures rarely exceeding 350°C, whereas fast pyrolysis typically employs temperatures of at least 450°C and processes relatively dry feedstock. Direct liquefaction is performed at very high pressures, approaching 200 atm, whereas fast pyrolysis is typically an atmospheric pressure process. Successful fast pyrolysis is based on very short reaction times whereas indirect liquefaction requires times as long as one hour to produce liquid products.

Like fast pyrolysis, direct liquefaction can produce a wide range of liquid products, including aliphatic and aromatic alcohols, phenols, hydrocarbons, substituted furans, and alicyclic compounds. Some researchers report that a significant portion of O_2 in the biomass is removed as CO_2 in the gaseous effluent. The resulting liquid has a heating value of 36 MJ/kg and the overall thermal efficiency, based on lower heating values of feedstock and product liquid, is 80–90%.

The process has not been as thoroughly investigated as fast pyrolysis. Certainly the prospect of continuous feeding of biomass into a high-pressure reaction vessel presents some difficulties to practical implementation.

7.5.3 Indirect Liquefaction

Indirect liquefaction produces liquid fuel by gasifying low-value organic materials to a mixture of H_2 and CO, known as *syngas*, followed by catalytic or biological synthesis to ethanol, methanol, or other chemical compounds. The process has been developed commercially for the conversion of coal into methanol or synthetic gasoline. The process can also employ biorenewable resources as feedstock, although the current low cost of fossil fuels has hampered the manufacture of biobased products by this method.

Two distinct approaches to converting syngas to chemicals are possible. The traditional approach involves a moderate-temperature, high-pressure catalytic chemical synthesis of methanol or synthetic gasoline. A more recently developed approach, yet to be commercially applied, employs microorganisms that are able to grow on one-carbon compounds to convert syngas into organic acids or alcohols.

7.5.3.1 Catalytic Approach

Methanol is formed by the exothermic reaction of one mole of CO with two moles of H_2:

$$CO + 2H_2 \rightarrow CH_3OH \tag{7.2}$$

Low temperatures and high pressures thermodynamically favor the production of methanol. Current commercial operations use a fixed catalytic bed operated at 250°C and 60–100 atm, with gas recycle to remove the large amount of heat released by this exothermic reaction. More recently, liquid-phase slurry reactors have been introduced to improve contact between syngas and catalyst, as well as to enhance the removal of heat from the reactor.

Biomass gasification does not yield pure syngas: in addition to CO and H_2 the gas mixture contains CO_2, CH_4, light hydrocarbons, and a variety of high-molecular weight organic compounds collectively known as tar. The high temperature reaction of this gas mixture with steam over a catalytic bed breaks down the organic compounds to CO and H_2. The resulting mixture of CO, CO_2, and H_2 may require additional processing to obtain the necessary H_2/CO ratio of 2.0 required for methanol synthesis, since gasification of most biomass yields substantially smaller ratios.

Hydrogen enrichment can be achieved by passing syngas and steam over a catalytic bed to promote the water-gas shift reaction:

$$CO + H_2O \rightarrow CO_2 + H_2 \tag{7.3}$$

Low temperatures thermodynamically favor this slightly exothermic reaction. To obtain satisfactory reaction rates, catalysts are employed in one or more fixed-bed reactors operated in the temperature range of 250–400°C.

Methanol is the starting point for additional chemical synthesis, including production of acetic acid, formaldehyde, and methyl tertiary-butyl ether (MTBE). Production of hydrocarbons from syngas can be directly accomplished by *Fisher-Tropsch (F-T) synthesis*, a procedure that reacts and polymerizes syngas to light hydrocarbon gases, paraffinic waxes, and alcohols. Additional processing can produce diesel fuel and gasoline. Both methanol synthesis and F-T synthesis require careful control of the H_2/CO ratio to satisfy the stoichiometry of the synthesis reactions as well as to avoid deposition of carbon, known as coking, on the catalysts.

7.5.3.2 Biocatalytic Approach

Certain microorganisms, known as *unicarbontrophs*, are able to grow on one-carbon compounds as the sole source of carbon and energy. These include some of the same microorganisms involved in anaerobic digestion of polysaccharides. Acetogens can convert CO or mixtures of CO and H_2 to fatty acids and, in some cases, to alcohols. These microorganisms are thought to involve a common reaction pathway that consumes CO and produces acetyl CoA as an intermediate. Methanogens can produce methane from mixtures of CO, CO_2, and H_2. Thus, syngas can be biologically converted to a variety of fuels and chemicals including ethanol, methane, acetic acid, butyric acid, and butanol.

Clostridium ljungdahli, a gram-positive, motile, rod-shaped anaerobic bacterium isolated from chicken waste, has received particular attention for its ability to co-metabolize CO and H_2 to form acetic acid (CH_3COOH) and ethanol (CH_3CH_2OH):

$$4CO + 2H_2O \rightarrow CH_3COOH + 2CO_2 \tag{7.4}$$
$$4H_2 + 2CO_2 \rightarrow CH_3COOH + 2H_2O \tag{7.5}$$
$$6CO + 3H_2O \rightarrow CH_3CH_2OH + 4CO_2 \tag{7.6}$$
$$6H_2 + 2CO_2 \rightarrow CH_3CH_2OH + 3H_2O \tag{7.7}$$

The wild-type strain of *C. ljundahlii* produces an ethanol-to-acetate ratio of only 0.05 with maximum ethanol concentration of 0.1 g/L. This ratio is very sensitive to acidity; decreasing the pH to 4.0 increases the ratio to 3.0. Other adjustments to the culture media and operating conditions nearly eliminate acetate production and increase ethanol concentration to 48 g/L after 25 days.

This gasification/fermentation route to biobased products from lignocellulosic feedstocks has several advantages compared to the hydrolytic/fermentation techniques described in previous sections of this chapter. Gasification allows very high conversion of feedstock to usable carbon compounds (approaching 100%), whereas hydrolysis only recovers about half the lignocellulose as fermentable sugars. Gasification yields a uniform product (a gaseous mixture of CO, CO_2, and H_2) regardless of the biomass feedstock employed whereas hydrolysis yields a product dependent on the content of cellulose, hemicellulose, and lignin in the feedstock. Finally, since the syngas is produced at high temperatures, gasification yields an inherently aseptic carbon supply for fermentation.

Biological processing of syngas has several advantages compared to chemical processing. The H_2/CO ratio is not critical to biological processing of syngas, thus making unnecessary the water-gas shift reaction to increase the hydrogen content of biomass-derived syngas. Whereas catalytic syngas reactors require high temperatures and pressures, biocatalysts operate near ambient temperature and pressure. Also, biocatalysts are typically more specific than inorganic catalysts.

Syngas fermentation faces several challenges before commercial adoption. Syngas bioreactors exhibit low-volumetric productivity due, in part, to low cell densities. Recycling or immobilization of cells in the bioreactor are possible solutions to this problem. Mass transfer of syngas into the liquid phase is also relatively slow. In commercial-scale aerobic fermentations, mass transfer of oxygen is generally the rate-limiting process. The problem will be exacerbated for syngas fermentations since the molar solubility of CO and H_2 are only 77% and 65% of that of oxygen, respectively. Dispersion of syngas into microbubbles of 50 μm diameter will be important to successful design of multi-phase bioreactors.

7.6 Additional Processing to Value-added Chemicals

The key to commercially successful chemical synthesis is a ready supply of inexpensive building blocks—chemical structures that are easily manipulated into the desired chemical compounds. The building blocks for the petrochemical industry include syngas, ethylene, propylene, butadiene, and a mixture of monocyclic aromatic hydrocarbons (principally benzene, toluene, and xylene) known as BTX. Syngas and the light alkenes are the starting points for alcohols, glycols, aldehydes, ketones, chlorinated hydrocarbons, esters, and ethers, as well as a variety of polymers. BTX, derived from naphthas in refineries, is the starting point for the hundreds of aromatic compounds used in the modern world.

A simple substitution of building blocks derived from biorenewable resources is not possible, because polysaccharides are polymers in which each carbon atom is bound to an oxygen atom. Carbon-oxygen bond scission could be accomplished in a number of ways to obtain the building blocks of the petrochemical industry. As previously described, biomass gasification yields syngas that can be catalytically converted to alkenes by the Fisher-Tropsch process. Alternatively, ethylene can be obtained from dehydration of ethanol obtained from fermentation of monosaccharides:

$$CH_3CH_2OH \rightarrow C_2H_4 + H_2O \qquad (7.8)$$

Fast pyrolysis (reactor temperatures and residence times of 400–500°C and 0.03 to 1.5 s, respectively) yields light alkenes such as ethylene (C_2H_4) and propylene (C_3H_6) but at relatively low yields, with other hydrocarbons and syngas as co-products.

The BTX fraction from petroleum refining, used as building blocks for aromatic compounds, has no analogue in thermal processing of biomass since only small quantities of monocyclic aromatic hydrocarbons are formed. Instead, pyrolysis of cellulose and hemicellulose yields anhydrosugars and their dehydration products, low-molecular weight oxygenates, and alicyclic compounds (that is, organic compounds that contain rings but are not aromatic). Although lignin has highly aromatic structures, it pyrolyzes to phenolic oligomers instead of the monocyclic aromatic hydrocarbons found in BTX.

Addition of hydrogen and catalysts during fast pyrolysis (hydropyrolysis) increases the yield of hydrocarbons, although a substantial quantity of oxygenated compounds remains. Similarly, hydrotreating of pyrolysis liquids or lignin obtained from fractionation of lignocellulose will remove oxygen as water and yield hydrocarbon liquids with higher hydrogen-to-carbon ratios than the feedstock. Deoxygenation without hydrogen is possible through catalytic vapor cracking, a procedure that accomplishes simultaneous dehydration and decarboxylation over acidic zeolite catalysts. At 450°C and atmospheric pressure, the zeolite catalysts are able to reject oxygen as H_2O, CO_2, and CO while producing either alkenes or aromatic hydrocarbons.

Although conversion of polysaccharides to petrochemical building blocks is technically feasible, it places biorenewable resources in direct competition with petroleum resources. This approach to introducing biobased products into the economy is problematic because of the relatively low prices of petroleum resources. Furthermore, since almost half of biomass by weight is oxygen, conversion to hydrocarbons would limit yields to 50% at best. A more viable approach may be the development of oxygenated organic compounds as building blocks for production of fuels and commodity chemicals.

The building blocks for biobased products include monosaccharides, lipids, the organic acids and alcohols from direct fermentation of monosaccharides, and the products of fast pyrolysis, such as levoglucosan. Both chemical and biological transformations of these building blocks can yield value-added products.

7.6.1 Chemical Conversions

The pentahydric alcohol known as xylitol is an important sweetener produced from xylose that has been recovered from the hemicellulose fraction of birch wood. Acid hydrolysis of the finely milled wood yields the five-carbon sugar xylose and oligomers, both of which are removed by a hot water wash. Hydrolysis of this stream yields additional xylose, which is separated from other pentoses. Combination of xylose with hydrogen in the presence of a nickel catalyst hydrogenates the aldehyde group of xylose to yield xylitol. Processes have also been patented for extracting the necessary xylose from corncobs and bagasse.

Furfural, a commercially important solvent used in the manufacture of pesticides, synthetic resins, and nylon, is produced by a series of acidic dehydrations of pentoses obtained from hemicellulose-rich agricultural materials such as corncobs, oat hulls, and peanut shells. Although classified as secondary processing of biorenewable resources, in practice the dehydration of pentose occurs simultaneously with the hydrolysis of hemicellulose in a single reactor known as a digester. Sulfuric acid and biomass are blended and reacted with steam in the digestor. Furfural, which boils at 160°C, appears as vapor, and exits with steam leaving the reactor. Distillation yields furfural and a wastewater stream containing about 1% acetic acid.

Furfural, illustrated in Fig. 7.9a, is a member of the *furan compounds*, the heterocyclic aromatic series characterized by a ring structure composed of one oxygen atom and four carbon atoms. The reactivity of the furan ring allows a wide range of synthesis options. Furfural is converted to furan (Fig. 7.9b), the simplest member of this series, by catalytic decarbonylation, or to furfuryl alcohol (Fig. 7.9c) by hydrogenation of the aldehyde group. Furan can be converted to tetrahydrofuran (Fig. 7.9d)

(a)   (b)   (c)   (d)   (e)

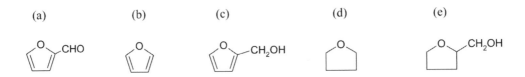

FIG. 7.9. Furan compounds synthesized from pentose: (a) furfural, (b) furan, (c) furfuryl alcohol, (d) tetrahydrofuran, (e) tetrahydrofurfuryl alcohol.

by catalytic hydrogenation. Furfuryl alcohol can be hydrogenated to tetrahydrofurfuryl alcohol (Fig. 7.9e). From tetrahydrofuran and tetrahydrofurfuryl alcohol a wide variety of derivatives can be obtained. The ring cleavage chemistry of furan suggests the basis for synthesis of a variety of other chemicals as well, including *n*-butanol, 1,3-butadiene, styrene, adipic acid, maleic anhydride, 1,4-butanediol, and *γ*-butyrolactone. Both furan and furfural have been converted to maleic anhydride. Levulinic acid can be produced by either acid treatment of tetrahydrofurfuryl alcohol or by acid treatment of hexoses (with the co-production of formic acid).

The fermentation product 2,3-butanediol can be dehydrated to methyl ethyl ketone (MEK) and used as an octane booster for gasoline. MEK can also be further dehydrated to 1,3-butadiene and dimerized to styrene. Thus, 2–3-butanediol can be the starting point for production of fuels, polymers, and antifreeze.

The hexoses yield a wide range of esters, ethers, and anhydro derivatives, and they undergo numerous rearrangement, substitution, isomerization, cyclodehydration, and reduction-oxidation reactions.

7.6.2 Biological Conversions

Biological processing employs enzymes to catalyze chemical reactions in highly selective manners. Enzymes are classified as hydrolases, which catalyze hydrolysis reactions; isomerases, which catalyze the transfer of groups; oxidoreductases, which promote oxidation-reduction reactions; lyases, which catalyze the removal or addition of groups to double bonds; and ligases, which join molecules at the expense of high-energy bonds. Careful design of bioreactors allows several enzymes to perform within a single reactor producing the same product as a more complicated, multi-step thermochemical conversion system.

Strains from several genera of bacteria, including *Klebsiella, Citrobacter,* and *Clostridium,* can anaerobically convert glycerol to 1,3-propanediol. The process involves two competing metabolic pathways. An oxidative pathway dehydrogenates glycerol to pyruvate. A parallel, reductive pathway dehydrates glycerol to 1,3-propanediol. Pyruvate is ultimately converted to one of the following: ethanol, acetate, butyrate, or 2,3-butanediol. Maximum yields are estimated to be 0.72 mol/mol for glycerol as the sole carbon source and 1.0 mol/mol for fermentation with co-substrates.

Further Reading

Conversion of Sugars and Starches

Blanchard, P. H. 1992. *Technology of Corn Wet Milling and Associated Processes*. New York: Elsevier.

Watson, S. A. and P. E. Ramstad, eds. 1987. *Corn: Chemistry and Technology*. St. Paul, Minnesota: American Association of Cereal Chemists.

Wayman, M. and S. R. Parekh. 1990. "Cereal grains." Chap. 4 in *Biotechnology of Biomass Conversion: Fuels and Chemicals from Renewable Resources*. Philadelphia: Open University Press.

Conversion of Lignocellulosic Feedstocks

Lynd, L. R. 1996. "Overview and Evaluation of Fuel Ethanol from Cellulosic Biomass: Technology, Economics, the Environment, and Policy." *Annual Rev. Energy Environ.* 21:403–465.

Wayman, M. and S. R. Parekh. 1990. "Ethanol from Wood and Other Cellulosics." Chap. 16 in *Biotechnology of Biomass Conversion: Fuels and Chemicals from Renewable Resources*. Philadelphia: Open University Press.

Lipid Extractives

Klass, D. L. 1998. "Natural Biochemical Liquefaction." Chap. 10 in *Biomass for Renewable Energy, Fuels, and Chemicals*. San Diego: Academic Press.

Thermochemical Conversion

Beenackers, A. A. C. M. and W. P. M. Van Swaaij. 1984. "Methanol from Wood: Process Principles and Technologies for Producing Methanol from Biomass." *Int. J. Solar Energy* 2: 349–367.

Bridgwater, A., S. Czernik, J. Diebold, D. Meier, A. Oasmaa, C. Peacocke, J. Piskorz and D. Radlein. 1999. *Fast Pyrolysis of Biomass: A Handbook*. Newbury, United Kingdom: CPL Press.

Mills, G. A. 1993. *Fuel* 73:1243–1279.

Scott, D. S., P. Majerski, J. Piskorz and D. Radlein. 1999. "A Second Look at Fast Pyrolysis of Biomass —the RTI Process." *J. Analytical & Applied Pyrolysis* 51:23–37.

Worden, R. M., M. D. Bredwell and A. J. Grethlein. 1997. "Engineering Issues in Synthesis-gas Fermentations." In *Fuels and Chemicals from Biomass*, B. C. Saha and J. Woodward, eds. ACS Symposium Series 666. Washington, D. C.: American Chemical Society.

Additional Processing

Grohmann, K., C. E. Wyman and M. E. Himmel. 1992. "Potential for Fuels from Biomass and Wastes." *Materials and Chemicals from Biomass*. Washington, D. C.: American Chemical Society.

Klass, D. L. 1998. "Organic Commodity Chemicals from Biomass." Chap. 13 in *Biomass for Renewable Energy, Fuels, and Chemicals*. San Diego: Academic Press.

Problems

7.1 Fermentation technologies that use lignocellulose as feedstock rather than starch represent a more efficient use of agricultural crops. Calculate the energy efficiency of converting switchgrass to ethanol and compare it to the conversion of corn grain to ethanol. Note that essentially the whole switchgrass crop is lignocellulose, whereas only a fraction of the corn plant is corn grain.

7.2 A 2000 barrel-per-day ethanol plant using Jerusalem artichokes as a source of sugar for fermentation is proposed.
   (a) Determine the daily feedstock requirement of raw Jerusalem artichokes for the plant.
   (b) Determine the number of hectares that must annually be planted to provide feedstock to the plant.

7.3 Determine the theoretical yield (mass basis) of ethanol from the following feedstocks:
   (a) Starch.
   (b) Lignocellulose, assuming only the cellulose component can be hydrolyzed and fermented.
   (c) Lignocellulose, assuming both the cellulose and hemicellulose components can be hydrolyzed and fermented.

7.4 Describe the major differences between the operation and output of dry-milling and wet-milling plants. What are the advantages of each?

7.5 Describe the steps required to liberate monosaccharides from lignocellulose. Calculate the approximate yield (wt-%) of glucose from lignocellulose.

7.6 Consider the conversion of producer gas from an oxygen-blown downdraft gasifier into ethanol. Estimate the amount of producer gas (m^3/day) required to yield 20,000 L/day of ethanol.

Processing of Biorenewable Resources into Natural Fibers

8.1 Introduction

Plant fibers are long, hollow cells that provide structural support and/or conduct water and nutrients through the plant. The cell walls of fibers consist primarily of lignocellulose. The previous chapter considered how to depolymerize this material into simple molecular building blocks. This chapter is devoted to processes that liberate fibers for the manufacture of biocomposites, paper, and other cellulose-based products.

The disintegration of plant material to recover fibers is known as *pulping*, a process that can be accomplished mechanically or chemically. *Mechanical pulping*, which physically separates the lignocellulosic fibers from one another, has yields of 91–98%. Mechanical pulp produces smooth printing surfaces of good opacity and has traditionally been employed in the production of newsprint. Mechanical pulp usually needs to be supplemented with chemical pulp, about 5%, to provide the strength required to survive modern high-speed printing presses.

Chemical pulping dissolves lignin, an adhesive in cell walls, thus liberating cellulosic fibers. The advantage of chemical pulping is the production of cellulosic fibers, which are an order of magnitude stronger than the lignocellulosic fibers produced by mechanical pulping. However, mechanical pulping also dissolves hemicellulose, with the result that chemical pulping yields are only 35–60%. Chemical pulping is used for the production of paper for packing, writing, hygienic products, and for cellulose derivatives and it accounts for 70% of the total worldwide production of pulp.

In principle, both woody and herbaceous plant materials can be used for the production of pulp. Indeed, agricultural residues were a major source of pulp in the United States and Europe until the 1920s when issues of quality and supply began to favor woody biomass. Today, over 90% of pulp and paper is produced from woody biomass.

Although the lengths of fibers from herbaceous materials are comparable to those in hardwoods, they are typically shorter than the fibers obtained from softwoods; thus paper made from herbaceous plant fibers has relatively low tear strength, an undesirable characteristic for many applications. Herbaceous materials also contain about 30 wt-% of pith or parenchyma cells and about 5 wt-% of dense epidermis material. Neither of these is fibrous in nature and when left in the pulp, they have deleterious effects on paper quality.

Another factor, possibly even more important to the demise of herbaceous-biomass pulping, were the increased labor costs associated with securing supplies of agricultural residues. These materials have very low bulk density, resulting in a tremendous volume to handle, and they rapidly degrade when exposed to the weather, requiring special provisions for long-term storage.

Depithing processes, described in a subsequent section, can substantially overcome many of the fiber quality issues, while modern methods of collecting, storing, and handling of herbaceous biomass, previously described in Chapter 4, should help overcome supply problems. Certainly, a substantial volume of agricultural residues goes unused in the world today.

The following sections describe various aspects of fiber processing including mechanical-pulping, chemical-pulping, and depithing operations.

8.2 Mechanical Pulping

Germany introduced *groundwood pulping* in 1840 as a source of fibers to replace increasingly scarce rags used in the production of paper. In this process, wood is mechanically ground by large stones. The pulp is then screened to remove knots and other large pieces of wood followed by washing and bleaching in preparation for papermaking. *Thermomechanical pulping* was developed in the 1930s. In this process, wood chips are ground under pressure, producing heat that weakens the lignin in the wood chips and makes the separation of fibers easier. Thermomechanical processing yields relatively cheap fibers suitable for newsprint. *Refiner mechanical pulping* was developed in the 1960s. In this process wood chips are passed repeatedly through a series of rotating discs to remove fibers.

Although mechanical pulping provides high yields of fiber, it is energy intensive and does not remove lignin, which degrades paper strength and is responsible for darkening of newsprint with time.

8.3 Chemical Pulping

The two principal chemical pulping processes are the *sulfite process* and the *sulfate process* (also known as the *kraft process*). Until recently, the sulfite process was preferred because it produced a much lighter, easier to bleach pulp. However, the environmental impact of the sulfite process is greater than the kraft process, which has well-established methods for recovering processing chemicals and utilizing lignin that would otherwise be discharged to the environment. Since the 1950s, the kraft process has gradually replaced the sulfite process, and today accounts for over 80% of chemical pulp. Nevertheless, the sulfite process presents some useful lessons in co-product utilization that will be described here.

8.3.1 Sulfite Pulping

The sulfite process is illustrated in the flow chart of Fig. 8.1. The sulfite-pulping liquor is an acid solution of bisulfite of calcium, magnesium, sodium, or ammonia that is added to wood chips in a digester operated at 160°C. In about two hours, the lignin is sulfonated and the salts of the sulfonic acid are dissolved. Hemicellulose is hydrolyzed to pentoses and hexoses, depending on the chemical composition of the hemicellulose. The backbone of hemicellulose from hardwoods and herbaceous materials is primarily xylan, which yields xylose upon hydrolysis. The hemicellulose from softwoods is predominately mannon, which yields mannose,

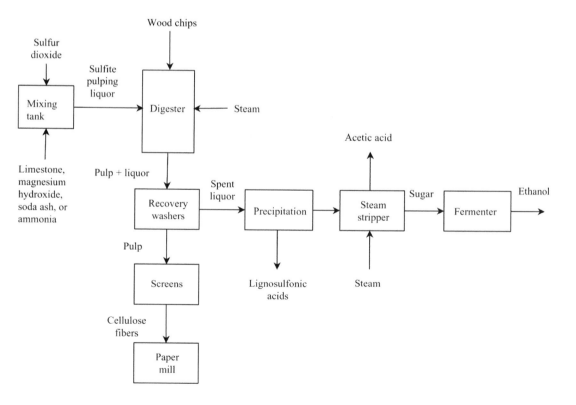

FIG. 8.1.   Flow chart of sulfite pulping.

an epimer of glucose, upon hydrolysis. In addition, the liquor from pulping hardwood contains almost twice as much acetic acid as the liquor from pulping softwood due to the higher acetyl content of hardwood hemicellulose. The sulfite process has not been generally employed for herbaceous fibers because of inferior yields and fiber strengths compared to the kraft process.

The pulp of cellulosic fibers is filtered from the spent liquor and then further processed to paper or cellulose-derivatives. The spent liquor contains carbohydrates and lignin, as well as pulping chemicals that should be treated as by-products rather than as a waste stream discharged to the environment. The sulfonic acids of lignin, which are highly condensed, high molecular-weight compounds, can be precipitated from the spent liquor and sold. They can be used as a binder in the preparation of fertilizers, animal feed, and fuel pellets; as a glue substitute and drilling-mud additive; and as a phenol substitute in the preparation of phenol-formaldehyde resins. Oxidation of lignosulfonic acids by air in strongly alkaline solution yields vanillin, a common flavoring agent and the raw material for the production of L-dopa, a pharmaceutical used for treatment of Parkinson's disease.

Until recently, it was common to ferment the hexoses in spent liquor from softwood pulping to ethanol. Although phenol from lignin fragments inhibits fermentation, yeast was adapted to tolerate its presence in spent pulping liquor. Scandinavian pulping mills, in particular, generated ethanol for use as a 25% blend with gasoline although the practice was phased out in 1958. Fermentation of spent liquor from hardwood pulping was not widely practiced for two reasons: the high acetic acid concentrations are toxic to most microorganisms, and xylose, the predominant sugar, is not readily fermented by common yeast. Steam stripping can reduce the concentration of acetic acid to less than 0.5 g/L, an acceptable level for fermentation. In recent years, good progress has been made in identifying and developing pentose-fermenting microorganisms. Alternatively, the xylose could be dehydrated to furfural, as described in Chapter 7.

8.3.2 Kraft Pulping

Kraft pulping was developed in Germany near the end of the nineteenth century in an effort to improve upon an early pulping process known as *the soda process,* or *alkaline pulping,* which used sodium hydroxide (NaOH) to pulp wood. The soda process could not compete with the sulfite process due to the relatively high cost of NaOH. The key innovation in kraft pulping was the replacement of sodium carbonate as make-up chemical with inexpensive sodium sulfate (Na_2SO_4). Upon heating, Na_2SO_4 is converted to sodium sulfide, which in turn reacts with water to form NaOH and sodium hydrosulfide (NaSH). The process produces a strong pulp, which was the source of its name, kraft being the German word for "strong." It is also known as the *sulfate process* for the compound used as make-up chemical. The kraft process was quickly adopted in the United States where it allowed com-

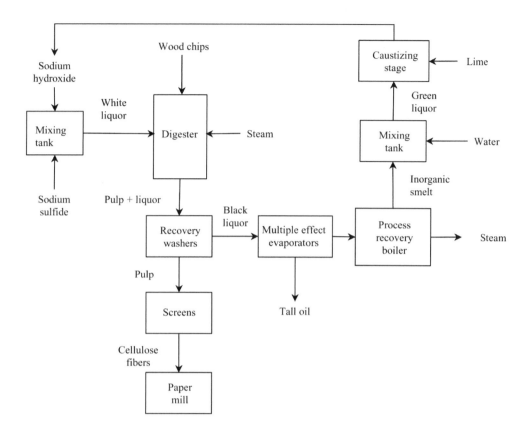

FIG. 8.2.   Flow chart of kraft pulping.

mercial pulping of Southern pines, which do not fare well in the sulfite process due to their high resin content.

The kraft process is illustrated in the flow chart of Fig. 8.2. Biomass is first steamed before being continuously added to a digester. Pulping is performed with a hot mixture of NaOH and sodium sulfide (Na_2S) known as *white liquor*. Roughly half of the biomass is degraded and dissolved to form a mixture of lignin, polysaccharides, and a small fraction of extractives. A substantial amount of the hemicellulose fraction of the wood is converted to hydroxy carboxylic acids. The lignin is degraded to a complex mixture of compounds. The soluble fraction is known as *kraft lignin* but has little resemblance to the native lignin in the wood.

The pulp of cellulosic fibers is washed to remove the spent liquor, known as *black liquor*, and screened to remove knots. Additional processing of the pulp may include oxygen delignification to further remove lignin binding fibers together and bleaching to brighten the fibers.

The black liquor is concentrated to 65–80% solids content with *multiple-effect evaporators*. Extractives present in the raw biomass and liberated during pulping can be removed during this stage. The resulting *tall oil* contains soaps formed from the saponification of fatty acid glycerides, and the esters of resin acids, especially in the pulping of softwoods. Acidification of these soaps yields *crude tall oil,* which contains 35% resin acids, 30% fatty acids, and 35% unsaponifiable terpenes and terpenoids. The latter resembles terpene hydrocarbons but with various oxygenated functional groups attached.

The black liquor also contains degraded polysaccharides and lignin. The polymeric lignin fraction can be precipitated by acidification, although kraft lignin has only limited applications compared to lignosulfonates obtained from sulfite pulping. Acid-precipitated kraft lignin had been used as an adhesive in panel boards. More generally, it has been employed as boiler fuel because of its high heating value compared to raw wood. The degraded polysaccharides are typically used as boiler fuel too, although their relatively low heating value suggests that higher-value applications should be explored.

The black liquor, containing degraded polysaccharides and lignin, is burned in a process-recovery boiler, which generates steam and recovers pulping chemicals. Combustion of black liquor produces an inorganic smelt of sodium carbonate (Na_2CO_3) and Na_2S with a small amount of sodium sulfate (Na_2SO_4). The smelt is dissolved in water to form *green liquor* that then reacts with lime (CaO) to convert Na_2CO_3 into NaOH and regenerate the original white liquor in the causticizing stage. Due to incomplete reaction in the recovery cycle, the white liquor also contains Na_2CO_3 and sodium salts of oxidized sulfur-containing anions. The use of the process-recovery boiler is integral to the economic operation of a kraft-pulping mill.

Herbaceous materials obtained from agricultural residues, including wheat and rice straw and bagasse from processed sugarcane, have relatively low lignin content, often only one-half that of woody material. Thus, chemicals more easily penetrate their plant cell walls, and pulping of herbaceous materials may be accomplished in only one-fourth to one-third the time required for woody materials. In fact, pulping is so rapid that there is no advantage in using sodium sulfide as part of the white liquor prepared for the kraft-pulping process. Instead, the simpler soda process, using only sodium hydroxide, is often used to pulp herbaceous biomass.

The pulping of agricultural residues is complicated by the large quantities of ash present, particularly silica, compared with woody materials. This silica appears in the black liquor of kraft or soda processes, where it creates severe problems in recovery operations, especially in the formation of silica scale on heat-transfer surfaces of evaporators and boilers.

Desilication of the black liquor can be accomplished with either lime treatment or by pH adjustment. Lime precipitates silica as calcium silicates by the reaction:

$$Na_2SiO_3 + CaO + H_2O \rightarrow CaSiO_3 + NaOH \qquad (8.1)$$

The calcium silicate precipitates from solution and is removed by filtration. Because this reaction also precipitates out sodium carbonate and sodium sulphate, pH adjustment by passing CO_2 in the form of flue gas through the black liquor is considered the best approach to desilication. This method precipitates SiO_2, which is removed by band filters. Alkali consumption is economized by vacuum stripping the CO_2 from the desilicated liquor, decomposing the bicarbonates formed.

8.4 Depithing

The stalks of some herbaceous plants, such as sugar cane and corn, contain 25–35% on a dry weight basis of parenchyma cells (*pith*), which do not have the characteristics of fiber. Pith, containing a high portion of hemicellulose, is readily penetrated and reacted by pulping chemicals; thus, pith consumes a large fraction of the chemicals while yielding little useable pulp. Furthermore, the outer surface of these stalks consists of tough, waxy epidermis cells that are resistant to pulping chemicals. Representing about 5% of the dry weight of stalks, epidermis cells leave troublesome residues in the finished pulp, usually appearing as dark specks. *Depithing* is the process of removing parenchyma cells from herbaceous plant stalks. Many types of depithing operations also remove epidermis cells, producing a pulp suitable for paper products that do not require high tear strengths.

Sugar cane bagasse, the fibrous stalk residue remaining after pressing the cane for sugar, has been widely explored as a source of pulped fibers. The following depithing operations were developed for bagasse, although they could equally apply to other herbaceous plant stalks.

Approximately 50% of the dry weight of bagasse stalks consists of high-quality fiber bundles concentrated in the hard, dense rind of the stalk. These fiber bundles are separated relatively easily from the pith tissue in which they are embedded. An additional 15% of the stalk is made up of shorter, less-resistant fiber bundles scattered throughout the interior pith portion. These bundles are more easily dissociated into individual fibers than those found in the rind. Separation of these two kinds of fibers during depithing operations is not generally considered economical.

Depithing can be classified into three categories: dry depithing, moist depithing, and wet depithing. *Dry depithing* separates the pith from bagasse after it has been dried to 10–25% moisture content. Although widely used prior to 1960, the method was inadequate for production of high-quality pulp. Other depithing methods are more appropriate for modern storage and processing of bagasse. *Moist* or *humid depithing* employs bagasse "as received" from a sugar mill, which usually has a moisture content of 50%. This operation is the most economical of the three, but does not achieve fibers as clean and uniform as can be obtained by *wet depithing*. This third type of depithing operation employs a slurry of water and bagasse to improve depithing. Wet depithing also has the advantage of removing epidermis cells as well as pith if stalks are well broken up in the milling process. However, wet depithing of raw bagasse generates a large amount of wet pith, which is difficult to handle.

Many commercial operations employ a two-stage depithing operation consisting of moist depithing, economically removing most of the pith, followed by wet depithing, removing the residual pith as well as epidermal cells from the fiber. An example of such a two-stage operation is illustrated in Fig. 8.3. As-received bagasse is fed to a depithing machine consisting of a rotor to which are attached swinging, or rigid, hammers. The mechanical action rubs or breaks loose the pith from the fiber, which is flung radially outward through screens or perforated plates enclosing the rotor. Fibers that are too large to pass through the screens fall downward where they are collected separately. If this process is performed at the sugar mill, the pith removed, representing 30–40% of the weight of bagasse, can be employed as fuel in the mill's boiler. The cost of transporting the resulting fiber to a pulping mill is substantially less than for transporting the raw bagasse.

The second stage of depithing occurs at the pulping mill. Bagasse and water are continuously added to a *hydrapulper*, which thoroughly wets the bagasse and helps break loose dirt and pith. The resulting slurry, which contains 2.5–3% solids, is

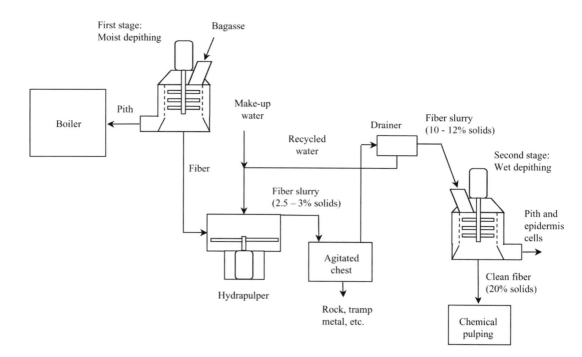

FIG. 8.3. Flow chart of two-stage depithing operation.

pumped to agitated chests where rock, tramp metal, and other heavy material set-
tle out. Since wet depithing machines are designed to operate with slurry of 10–
12% solids content, the slurry passes through a dewatering device such as a rotary
drum, vibrating screen, drag drainer conveyor, or a screw-type drainer conveyor.
The drainers remove a considerable amount of dirt and some pith cells from the
slurry, which is then pumped to the wet depither. The *wet depithing machine*, like
moist depithing machines, consists of hammers attached to a rotor enclosed within
a perforated drum. Pith and epidermal cells are small enough to pass through per-
forations, while the clean fiber slurry, with solids-loading increased to about 20%,
flows out the bottom. A screw feeder conveys the fiber slurry to the chemical pulping digester. The wet pith leaving the machine has a solids content of only about
1%. In some parts of the world this waste stream is used to irrigate cane fields.
Otherwise, wet pith must be dewatered to at least 15% for landfill, or to approximately 50% for use as boiler fuel.

Further Reading

Woody-biomass Pulping

Alén, R. 2000. "Basic Chemistry of Wood Delignification." *Forest Products Chemistry*. Per Stenius, ed.
   Helsinki: Fapet Oy.
Wayman, M. and S. Parekh. 1990. "Wood." Chapter 6 in *Biotechnology of Biomass Conversion: Fuels and
   Chemicals from Renewable Resources*. Philadelphia: Open University Press.

Herbaceous-biomass Pulping

Hamilton, F. and B. Leopold, eds. 1987. *Pulp and Paper Manufacture: Volume 3. Secondary Fibers and
   Non-Wood Pulping*. Atlanta: Joint Textbook Committee of the Pulp and Paper Industry.
Rowell, R. M. 1992. "Opportunities for Lignocellulosic Materials and Composites." *Emerging Technolo-
   gies for Material and Chemicals from Biomass*, R. M. Rowell, T. P. Schultz and R. Narayan, eds. ACS
   Symp. Series 476. Washington, D.C.: American Chemical Society.
Rowell, R. M., R. A. Young and J. R. Rowell. 1997. *Paper and Composites from Agro-Based Resources*.
   CRC Press.

Problems

8.1 Indicate the relative advantages and disadvantages of mechanical and chemical pulping.

8.2 Why does wood pulp dominate the markets for pulp and paper compared to
pulp from agricultural materials such as bagasse and corn stover?

8.3 Indicate the relative advantages and disadvantages of the sulfite and kraft pulping processes. What factor accounts for the dominance of kraft pulping in today's pulp and paper industry?

8.4 What is black liquor? What is its composition? What is the fate of this co-product in a pulp and paper mill?

8.5 Explain why depithing is required for commercial pulping of fibers from herbaceous feedstocks.

Environmental Impact of the Bioeconomy

9.1 Introduction

Much has been made of the negative environmental impact of using fossil fuels. The assumption has often been made that anything that reduces use of fossil fuels will automatically benefit the environment. The reality is much more complicated —every technology introduces both benefits and costs whose net impact depends on what is most valued by society. Exploitation of biorenewable resources is no exception. Agriculture has been blamed for desertification in Africa, while demand for fuel wood was clearly responsible for the deforestation of large parts of Europe.

This chapter explores some of the potential environmental impacts of using biorenewable resources in the production of fuels, chemicals, energy, and fibers. These impacts are not always readily apparent. For example, the role of chlorofluorocarbons, once commonly used as refrigerants and blowing agents in production of foams and insulation, in atmospheric ozone depletion was not recognized until decades after their commercial development. Thus, there is a tendency to err on the side of caution in assessing potential negative impacts of new technologies on the environment. This chapter organizes the topic in terms of the broad areas of plant science, crop production, processing, and utilization of biobased products.

9.2 Plant Science

The potential impact of plant science on the environment has spawned one of the most contentious debates in modern society. The prospect of creating new plants by the application of biotechnology and releasing them to the biosphere or introducing them to the human food supply concerns many people. Particularly in Western nations, where people enjoy secure food supplies and have the leisure and financial resources to appreciate pristine natural environments, there has arisen the notion of "Frankenfood," a man-made plant with toxins that poison the food supply or that is so prolific that it upsets the balance of nature.

Application of biotechnology to plant science offers several opportunities for plant improvements, including resistance to herbicides, protection against insects and other pests, hardiness against frost, and the manufacture of valuable products in plants. The latter of these possibilities, many believe, is the key to future growth in the development of biobased products. The concerns about the environmental impact of transgenic crops must be addressed early in the development of these crops if this future is to be realized. Some of the most prominent environmental concerns about transgenic crops are illustrated in Fig. 9.1 and discussed in the following paragraphs.

As described in Chapter 4, the process of developing a transgenic plant requires selection of plant cells that have been successfully transformed; that is, those that have incorporated the transgene. This is commonly accomplished by inclusion of a selectable marker gene in the constructed transgene that is resistant to an herbicide or an antibiotic. Thus, the application of the herbicide or antibiotic to regenerated plants kills those that do not contain the transgene, allowing the transgenic plants to be identified and selected. The prospect of these resistant genes being transferred to wild plants by pollination, or to bacteria by the same kind of gene transfer between bacteria and plant cells that are frequently used to produce trans-

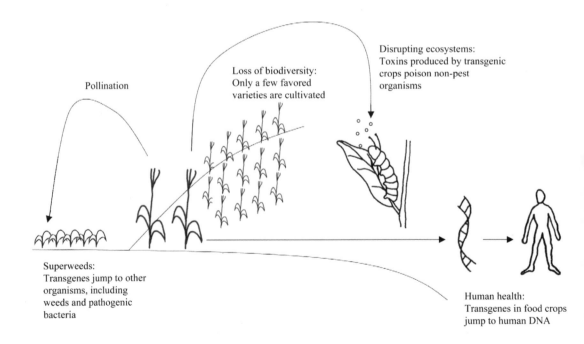

FIG. 9.1. Potential environmental impacts of plant science.

genic crops, is disturbing to many people. Transfer of herbicide resistance to weeds would be a serious threat to agriculture, while transfer of antibody resistance to pathogens would threaten human health.

Transfer of herbicide resistance is a distinct possibility if transgenic crops are grown in close proximity to closely related weed species. For example, gene movements between wild mustard and a crop of canola and between jointed goatgrass and a crop of wheat have been demonstrated, raising the prospects for "superweeds." The transfer of antibiotic resistance to pathogenic bacteria is an extremely unlikely event but cannot be completely ruled out. For these reasons, alternative selection methods, such as marker genes that cause a transformed plant to fluoresce when exposed to ultraviolet light, are under development. Of course, the transfer of genes that are specifically added to make a plant resistant to herbicides or to attack by insects cannot be prevented in this manner. In this case, more sophisticated approaches to preventing the creation of superweeds are required. For example, herbicide-resistant genes could be linked to other genes that are harmless to a crop but damaging to a weed that might incorporate the transgene.

Another concern is that parts of a transgene inserted into food crops might escape the digestive tract and be inserted into human chromosomes. A specific example cited by critics of transgenic crops is the *cauliflower mosaic virus (CaMV)*, often used as a promoter sequence in transgenic crops. (Recall that a promoter sequence is the on/off switch that controls where the plant gene will be expressed.) Research on plasmid DNA in rice is cited as evidence that the CaMV promoter can insert itself into strands of DNA. However, the chain of events necessary for this to occur in human DNA is unlikely and has never been observed. CaMV infects several common crop plants, including cauliflower, broccoli, cabbage, bok choy, and canola, and has been consumed by animals and humans for centuries with no documented negative effects on health.

One well-publicized concern is that transgenic crops might threaten other organisms in an ecosystem by introducing toxic substances. Researchers have documented in laboratory trials that pollen from Bt corn, one of the first commercial transgenic crops, is toxic to monarch butterfly larvae. Bt corn was genetically engineered to contain a substance toxic to natural predators of the corn plant, so the fact that Bt corn pollen was toxic to insects was not particularly surprising. However, this demonstration raised the possibility that pollen drifting from a cornfield to milkweed growing in adjacent fields would poison monarch butterfly larvae that feed on milkweed plants, their principal source of nourishment. Follow-up studies indicate that pollen drift is unlikely to produce toxic levels of Bt corn pollen in adjacent fields. The original laboratory work, though, does demonstrate the delicate interplay between introduced plants and their environment.

Another argument cautions that transgenic crops will ultimately reduce biodiversity as they replace traditional crops. Inevitably this will be true, since the history of agriculture even before transgenic crops might best be described as the adoption of a few superior varieties that replaced a wider selection of traditional crops. The solution to loss of biodiversity is not to limit choices in agriculture but rather to establish collections of rare and obsolete varieties to serve as gene banks.

In developing transgenic crops, risk assessment should evaluate the following questions regarding their release to the environment: Does the introduction of a transgenic plant with resistance to a particular pest or disease exacerbate the emergence of new pests or diseases that are worse than the original? If a transgenic trait is transferred to a wild species, does its geographical expansion represent a threat to biological diversity? Would adoption of stress-tolerant plants lead to significant destruction of natural ecosystems?

9.3 Production

Agriculture can have either positive or negative impacts on the environment depending upon how it is practiced. Early cultivation of cotton and tobacco in the southern United States was notorious for quickly depleting soil fertility, requiring frequent opening of new lands for production of these crops. As illustrated in Fig. 9.2, unsustainable agricultural practices result in loss of soil fertility, soil erosion,

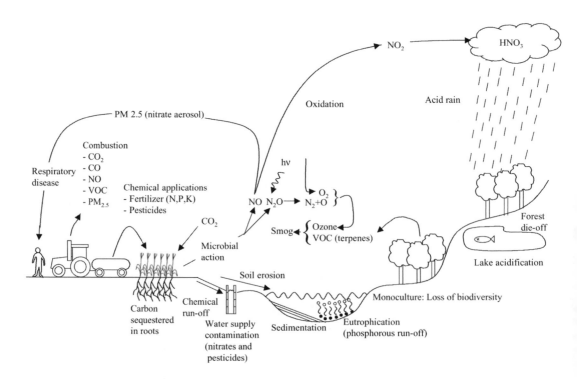

FIG. 9.2. Potential environmental impacts of crop production.

water pollution, air pollution, reduced biodiversity, and net positive emission of greenhouse gases from the soil to the atmosphere. Careful application of conservation principles can avoid or even reverse these impacts of agriculture.

9.3.1 Soil Fertility

Soil fertility is primarily a function of the amounts of organic carbon (also known as humus), nitrogen (N), phosphorous (P), and potassium (K) in the soil. The amount of organic carbon in the soil increases or decreases from year to year depending on tillage practices. The levels of inorganic nutrients (N-P-K) in the soils of natural ecosystems are determined by a balance between wind- and water-transport processes and uptake of those nutrients by standing biomass.

Organic carbon affects soil fertility in several ways including increasing water retention capacity and gas permeability, and improving the availability of inorganic nutrients to plant roots. Organic carbon is a natural by-product of the decay of plant material in oxygen-poor environments. This includes roots as well as above-ground stems and leaves that accumulate on the surface of soil as thatch or are submerged in waterlogged soils. Plant material exposed to air will eventually oxidize to CO_2, which is released to the atmosphere. Organic carbon accumulated to many meters of depth in the tall grass prairies of the midwestern United States in the period between the last Ice Age and the present era. This deep, black soil is the basis for the highly productive agriculture system in the United States. Ironically, the process of plowing these soils for agriculture exposed the buried carbon to oxidation, thus contributing substantially to the loss in fertility of these soils over the past 150 years. Modern conservation-tillage practices can reduce and even reverse these losses.

Intensive agriculture disrupts the balance of inorganic nutrients in the soil. The uptake of nitrogen, phosphorous, and potassium by standing biomass that is subsequently removed as a crop leads to depletion of these nutrients from the soil faster than the transport by wind or water can replace them. Consequently, inorganic nutrients must be periodically replaced by application of fertilizer. Cropping systems have widely varying fertilizer requirements, as indicated in Table 9.1. Corn, for example, is notable for its demand for nitrogen, requiring on the order of 135 kg/ha/yr compared to only 50–60 kg/ha/yr for herbaceous energy crops (HEC) and short rotation woody crops (SRWC). At the other extreme is soybean, requiring only 20 kg/ha/yr of nitrogen as a result of this plant's ability to synthesize nitrogen compounds from atmospheric nitrogen through the action of symbiotic bacteria in its roots. Not only is fertilizer application expensive and energy intensive (especially in the case of nitrogen), but it also impacts water pollution, as subsequently described.

Some dedicated energy crops have the potential for improving soil fertility. Those that are perennial crops, once established, do not require annual tilling of the soil, which contributes to the loss of soil carbon by both oxidation and erosion.

Table 9.1. Environmental impacts of different kinds of crops

Cropping System	Soil Erosion Rate (Mg ha^{-1} yr^{-1})	N-P-K Application Rate (kg ha^{-1} yr^{-1})	Herbicide Application Rate (kg ha^{-1} yr^{-1})	Insecticide Application Rate (kg ha^{-1} yr^{-1})
Annual crops				
Corn	21.8	135–60–80	3.06	0.38
Soybeans	40.9	20–45–70	1.83	0.16
Perennial crops				
HEC	0.2	50–60–60	0.25	0.02
SRWC	2.0	60–15–15	0.39	0.01

Source: W. G. Hohenstein and L. L. Wright, 1994, "Biomass Energy Production in the United States: An Overview," *Biomass and Bioenergy*, 6:161–173.

As Table 9.1 shows, these perennial crops also have relatively modest fertilizer requirements, especially compared to corn; thus, they deplete inorganic nutrients more slowly. In general, the establishment of dedicated energy crops in place of conventional row crops will improve soil fertility. However, this may not be true if the dedicated energy crops are established on fallow lands or pasture, especially in the early years of establishment.

9.3.2 Soil Erosion

Soil erosion is the transport of soil from one location to another, either by wind or water. In the process, the losses of organic carbon and inorganic nutrients from one location are gained by other locations; thus, erosion is not inherently bad. In practice, though, the net effect of soil erosion is to rob fertility from the lands under intensive agriculture, reducing crop productivity, and distributing soil and nutrients to unwanted locations, thus making them pollutants. Table 9.1 shows that cropping systems employing perennial crops have soil erosion rates that are one to two orders of magnitude less than for row crops such as corn and soybeans. In fact, during the first two years of establishment of perennial crops, erosion rates can be comparable to those for row crops, but these rates decrease dramatically after root systems develop and, in the case of woody crops, tree canopies close. Use of ground covers might further reduce erosion for dedicated energy crops.

9.3.3 Water Pollution

Water pollution arises from water-induced soil erosion as well as from leaching of chemicals (inorganic nutrients, herbicides, and insecticides) from soils. Transport of soil from cultivated fields into waterways is responsible for the notoriously muddy rivers and streams of the midwestern United States. Suspended solids can dramatically change aquatic ecosystems by, for example, driving out game fish that are then replaced by bottom feeders. Soils washed into reservoirs and lakes drop out of suspension to form sediments that seriously reduce the water-holding capacity of these impoundments, affecting both animal and human activities. Suspensions of soil can also facilitate transport of agricultural chemicals, as they can attach to individual soil particles.

Leaching of chemicals from soils is roughly proportional to their application rates to cropland. Although comprehensive data are not available, as little as 40% of nitrogen applied to an annual crop is incorporated into plant matter, the balance being volatilized or leached from the soil. Table 9.1 indicates that leaching of inorganic nutrients, herbicides, and insecticides can be expected to be considerably worse for row crops than perennial energy crops.

Ammonia, the usual form of nitrogen in fertilizer, is oxidized to nitrates when exposed to air. Nitrates are readily leached from soils and can appear in both well water and river water at concentrations exceeding the 10-ppm human health standard set by the U.S. Environmental Protection Agency (EPA). Human babies are susceptible to nitrate poisoning in the first few months of life when bacteria living in their digestive tracts convert nitrate to nitrite, which binds to hemoglobin and destroys its oxygen-carrying ability. Known as "blue baby" disease, a poisoned infant gradually suffocates. By the time a child reaches six months of age, stomach acids are sufficiently strong to kill the offending bacteria, and the disease is no longer a concern.

Phosphorous, which binds tightly to soil particles, is washed from fields as a result of soil erosion. Phosphorous represents a particular threat to aquatic ecosystems. In a process known as *eutrophication*, phosphorous promotes the rapid growth of algae near the surface of bodies of water. Eventually the algae die and their decomposition reduces dissolved oxygen to levels too low to support aquatic organisms and a general die-off occurs in the body of water. The problem can reappear periodically since phosphorous accumulates in the sediments of lakes and streams. Rapid run-off during a storm can stir up sediment and release phosphorous in another cycle of eutrophication.

Potassium in drinking water is not regulated. The World Health Organization does not consider this chemical to have any impact on drinking water quality or on human health.

Herbicides and insecticides are, by definition, toxic chemicals used to control plant and animal pests. Their environmental and health effects at low concentrations are not well known but they have been implicated as potential carcinogens and endocrine disrupters. They can be directly leached from the soil or transported by attachment to eroded soil particles. Most problems with pollution from pesticides arise from point-source emissions, such as occur from accidental spills or improper disposal of chemical containers. However, non-point-source pollution can result from heavy rainfalls that occur immediately after application of pesticides to a field. Both herbicides and insecticides have been detected year-round in unconfined aquifers at very high concentrations (hundreds of parts per million). The lower pesticide application rates and the lower soil erosion rates associated with perennial energy crops suggest that pesticide pollution will be considerably less than for conventional row crops.

9.3.4 Air Pollution

The contribution of crop production to air pollution is of two types: exhaust emissions from tractors and other production machinery, and dust and gas arising from tilled soils. Exhaust emissions from internal combustion engines include nitrogen oxides, CO_2, unburned hydrocarbons, and fine particulate matter. The impacts of these emissions on the environment are detailed in a subsequent section; however, the total emissions from production machinery are small and diffuse compared to those from the transportation and utility sectors of the economy.

Potentially more significant is the contribution of soil tillage to air pollution, which generates both particulate and gaseous pollutants. *Particulate pollutants* are classified as either *primary* or *secondary particulate matter (PM)*, depending upon its origin. Primary PM is emitted to the atmosphere as particles while secondary PM is frequently formed in the atmosphere by chemical reaction of gaseous pollutants. An example of primary PM is dust stirred up from dry soil when it is plowed. Primary PM produces relatively course particles, on the order of 10 μm in size.

A prominent instance of secondary PM is nitrate aerosols, which are formed by the oxidation of combustion-generated nitric oxide that has been emitted to the atmosphere. Secondary PM is produced as very fine particles, which, for regulatory purposes, is defined as being less than 2.5 μm in size. Such fine particle matter, designated as *PM 2.5*, is readily respired into the lungs but not out of them. PM 2.5 recently has been implicated in respiratory disease and is a regulated pollutant for some industries.

Wind-blown dust from agriculture can travel hundreds and even thousands of miles. Although a nuisance, this dust is mostly coarse PM, which is not thought to represent a health hazard. Careful soil management practices can mitigate nuisance dust from agriculture. However, agriculture can also produce secondary PM, which is mostly PM 2.5. Ammonia applied to soil as nitrogen fertilizer can escape as gas into the atmosphere where it reacts to form fine nitrate aerosol. Agriculture is not currently regulated for fine PM.

Several gaseous pollutants can arise from cultivation. Soil nitrogen, from either nitrogen-fixing bacteria or synthetic fertilizers, is converted to nitric oxide (NO) and nitrous oxide (N_2O) by microbial processes in wet, anaerobic soils. Methane, a greenhouse gas, is also produced by microbial processes in tilled soils; however, the vast majority of methane emissions come from animals and anaerobic digestion of animal wastes. Manure from livestock in the United States is estimated to emit about 3 million metric tons of methane annually and account for approximately 10% of the total U.S. methane emissions.

Nitric oxide released to the atmosphere is further oxidized to nitrogen dioxide (NO_2), which results in two pollution problems. As a strong oxidizing agent, NO_2 reacts to form nitric acid, which dissolves in water droplets to form *acid rain*. Environmental degradation from acid rain includes acidification of lakes, which kills marine organisms, and causes damage to forests and crops. Nitrogen dioxide is also critical to the formation of ground level ozone (O_3) in a two-step reaction mediated by sunlight:

$$NO_2 + h\nu \rightarrow NO + O \qquad (9.1)$$
$$O_2 + O \rightarrow O_3 \qquad (9.2)$$

Ground level ozone, also known as urban ozone, is a health hazard that has proven a serious concern in large cities where NO is formed by automobiles and power plants. It also reacts with volatile organic hydrocarbons emitted from automobiles to form *photochemical smog*, a mixture of aldehydes, acid gases, and peroxyacetal nitrate that forms a brown hazy smudge over many large cities. Smog is not only unsightly but irritates the eyes and lungs.

Although nitrogen oxide emissions in large urban areas are clearly a major problem, it is not known whether nitrogen oxide emissions from agricultural lands are a substantial problem since little information is available on the subject. It is expected, though, that emissions from dedicated energy crops will be less than from row crops but more than from temperate forests.

Nitrous oxide is relatively stable to chemical reaction compared to NO, but it is a strong greenhouse gas. The background level of N_2O in the atmosphere has been increasing during historical times. Some researchers attribute this trend to intensive agriculture. Its possible role in global warming is discussed in a later paragraph.

Volatile organic compounds (VOC) are also emitted from growing plants. Mostly terpenes, these molecules can react with nitrogen oxides in a process akin to smog formation in large cities to produce a bluish haze that gives the Smoky Mountains their name. This picturesque effect may be less appealing if anthropogenic emissions of nitrogen dioxides combine with biogenic emissions of VOCs in agricultural and pristine regions.

9.3.5 Biodiversity

Biodiversity describes an environment characterized by large numbers of different species of plants and animals. Agriculture traditionally has strived for just the opposite—a monoculture of one plant species where other plants and animals are considered pests. Production of dedicated energy crops has some prospects for improving biodiversity in the agricultural landscape. Certainly, the meadow-like

setting of fields planted to perennial grasses, and the forest-like setting of tree plantations, more closely resemble natural ecosystems than do row crops.

In fact, there are advantages in encouraging a certain degree of biodiversity in
dedicated feedstock supply systems. Multi-species production systems could
reduce the risks associated with pests. Adding nitrogen-fixing plants could reduce
fertilizer applications. Other plants might provide superior erosion control during
establishment.

9.3.6 Global Warming

Greenhouse gases are gases that interact with infrared radiation emitted from the
Earth in such a way as to trap this radiant energy in the Earth's atmosphere. Greenhouse gases are important in moderating the Earth's temperature, keeping it within
a zone that is conducive to life. The principal greenhouse gases are water, CO_2,
CH_4, N_2O, and chlorofluorocarbons. The latter arises solely from human activity,
while the others have both anthropogenic and biogenic origins. As the concentrations of these gases increase, the temperature of the Earth is expected to increase,
although the sensitivity of this temperature change to greenhouse gas concentrations is the subject of much debate.

What is known with certainty is CO_2 in the atmosphere has increased about
12% in the past fifty years, and average global temperature has been increasing
steadily for the last twenty years. Other greenhouse gases are also increasing,
although the CO_2 level is thought to be the most important determinant in global
temperature modulation. There is little debate that increases in atmospheric greenhouse gases have been strongly influenced by human activity. The burning of fossil fuels for heat and power is implicated in much of the anthropogenic emissions
but agriculture has also played a role. The opening of the prairies in North America 150 years ago exposed rich deposits of soil carbon to oxidation, and, more
recently, the wide-spread burning of rain forests in South America continues to
release large quantities of CO_2 into the atmosphere. The large stocks of cattle raised
today, and the cultivation of rice, are sources of CH_4 and N_2O.

Less certain is the magnitude of global climate change and the urgency for making changes in human activity. The historical and geological records provide only
limited information in this regard. Reliable temperature data have only been
recorded in very recent times. Various methods for estimating temperature and
atmospheric gas concentrations from the geological record suggest a positive cor-
relation between these parameters, but these are subject to considerable interpre-
tation. Increasingly sophisticated computer models are being developed to predict
long-term climatic changes, but they remain crude approximations of the real
world. Continued research is expected to provide increasingly reliable evidence on
how human activity affects global climate.

If society decides to reduce accumulation of greenhouse gases, then agriculture will have an important role to play. Although agriculture to date has generally been a net emitter of greenhouse gases, with proper practices it can not only drastically reduce greenhouse-gas emissions but also sequester carbon from the atmosphere. Methane absorbs infrared radiation about twenty-five times more efficiently than CO_2. Despite its relatively small concentration in the atmosphere, CH_4 is thought to be responsible for 12% of the global warming effect. Thus, one of the most obvious ways for agriculture to reduce greenhouse-gas emissions is to divert manure to covered anaerobic digesters where methane can be captured and used for heat and power. Although combustion of one mole of methane releases one mole of CO_2, the result is no net emission of greenhouse gas to the atmosphere since the manure was derived from plants whose origin was atmospheric CO_2.

Dedicated feedstock supply systems have a unique ability to sequester carbon from the atmosphere. As the plants grow, they absorb CO_2 from the atmosphere and convert it into carbohydrates, oils, or proteins. If the aboveground plant material is harvested and used for bioenergy or fuels, CO_2 is returned to the atmosphere with no net carbon emission. If the aboveground plant material is harvested and processed into biomaterials such as plastics or biocomposites, the associated carbon is sequestered for the life of the material, which might be as many as fifty years. Furthermore, the roots, which are underground plant material, contribute to soil humus, which could be sequestered for centuries.

The contribution of roots to sequestered carbon is not trivial; some perennial plants produce almost as much belowground biomass as aboveground biomass. Plantations of short-rotation woody crops might increase carbon inventories of soil by 30–40 Mg/ha over a period of 20–50 years. Maximum carbon inventories might take a century to be reached.

9.4 Processing

9.4.1 Heat and Power

The environmental performance of biorenewable resources for the production of heat and power is generally superior to that of fossil resources. Nevertheless, both kinds of resources contain various contaminants that contribute to air pollution when burned. These contaminants include nitrogen, sulfur, chlorine, and, especially for coal and municipal solid waste, heavy metals such as arsenic, cadmium, chromium, lead, and mercury. The concentrations of nitrogen, sulfur, and chlorine for different kinds of fuels are summarized in Table 9.2.

Among the fossil fuels only coal contains significant quantities of nitrogen, sulfur, and chlorine. The relatively small amounts of these contaminants in the raw

Table 9.2. Concentration of nitrogen, sulfur, and chlorine contaminants in various kinds of fuels

	Nitrogen (wt-%)	Sulfur (wt-%)	Chlorine (wt-%)
Gasoline	—	0.01	—
Bituminous coal	1–2	0.4–3	0.02–0.10
Refuse-derived fuel	0.6–0.8	0.3–0.5	0.5–0.6
Woody crops	0.06–0.6	0.01–0.02	0.01–0.10
Herbaceous crops	0.6–1.2	0.01–0.2	0.03–0.6

Sources: Various.

feedstocks for natural gas and petroleum-derived fuels are substantially removed during the preparation of commercial-grade fuel products (with the exception of diesel fuel for stationary power applications, which may contain up to 2.0 wt-% of sulfur). Coal also shows the highest concentrations of heavy metals, although diesel oil can contain substantial amounts of vanadium.

Biorenewable resources usually contain less than half the amount of nitrogen as does coal. Wood from traditional lumbering operations may have as little as 0.06 wt-% nitrogen, although nitrogen-fixing trees such as black locust and heavily fertilized short-rotation woody crops may contain ten times this amount of nitrogen. Agricultural residues from heavily fertilized herbaceous crops can contain as much as 1 wt-% nitrogen. Biorenewable resources may contain two orders of magnitude less sulfur as coal. On the other hand, biorenewable resources may contain an order of magnitude more chlorine than coal.

Refuse-derived fuel, obtained from municipal solid waste, is a mixture of fossil-resource-derived materials, such as plastic, and biorenewable resource-derived materials, such as paper. Accordingly, the amount of nitrogen, sulfur, and chlorine in refuse-derived fuel tends to be intermediate between fossil and biorenewable fuels. The levels of some heavy metals such as cadmium and mercury can be high if care is not taken to remove metal-bearing scrap from municipal solid waste.

The potential environmental impacts of processing biorenewable resources to heat and power are illustrated in Fig. 9.3. These include formation of acid rain, urban ozone, smog, and PM 2.5. *PM 10*, arising from ash and soot in flue gas, is also a source of air pollution. Incomplete combustion is a source of both VOC, often referred to as unburned hydrocarbons when arising from combustion, and CO. Water pollution from inadequate treatment of wastewater can also be a problem.

During combustion most fuel-bound nitrogen is oxidized to nitric oxide (NO), with smaller amounts of nitrogen dioxide (NO_2) and nitrous oxide (N_2O). These nitrogen oxides are collectively designated as NO_x. Gasification and pyrolysis convert fuel-bound nitrogen to ammonia (NH_3) and hydrogen cyanide (HCN). Unless the producer gas is chemically scrubbed, these nitrogen compounds are subsequently oxidized to NO_x when the producer gas is burned.

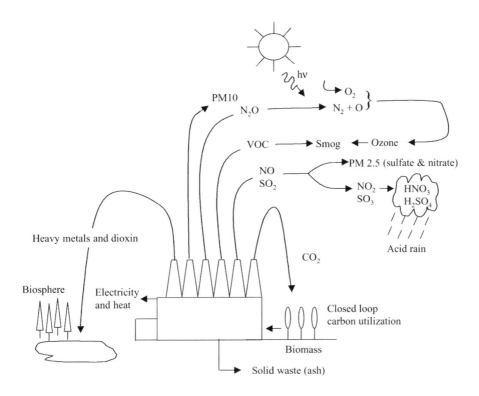

FIG. 9.3. Potential environmental impacts of processing biorenewable resources into heat and power.

Based on fuel nitrogen content, coal would appear to be the worst source of NO_x, followed by biorenewable sources, whereas natural gas and petroleum-based fuels would be expected to produce only minor amounts of NO_x. In fact, combustion of any type of fuel in air can produce large quantities of NO_x in considerable excess of the amount expected from fuel-bound nitrogen. The reason for this is that nitrogen and oxygen in air can react at high temperatures to produce NO:

$$N_2 + O_2 \leftrightarrow 2NO \tag{9.3}$$

Good combustion practice at one time dictated the use of high temperatures and high excess air (oxygen) levels to assure complete oxidation of carbonaceous fuels. However, these conditions also favor formation of NO. In fact, high levels of NO in flue gas were once used as an indicator of proper boiler operation. This "thermal NO_x" was the primary source of nitrogen compounds from power plants until environmental regulation in the late twentieth century forced boiler operators to adjust operating conditions to substantially eliminate it. Today, most nitrogen compounds from power plants are "fuel NO_x," and the amount emitted from any given boiler is proportional to the nitrogen content of the fuel. However, even fuel NO_x can be reduced well below the amount expected stoichiometrically by the use of various NO_x control technologies. These include staged combustion and fuel reburning, which exploit oxygen-starved conditions to reduce NO to N_2; *selective catalytic reduction (SCR)*, which injects amine or ammonia into flue gas to react with NO to form N_2; and *selective non-catalytic reduction (SNCR)*, which employs a catalyst to reduce NO. The control of NO_x from burning liquid fuels, which contain very little fuel-bound nitrogen, is discussed under the section on fuels and chemicals.

Sulfur in fuel is oxidized to SO_2 during combustion or released as H_2S during gasification. Unless chemically scrubbed from producer gas, H_2S will be oxidized to SO_2 when the producer gas is burned. Sulfur dioxide is further oxidized in the atmosphere to form sulfuric acid, which can form acidic water droplets or sulfate aerosols. Thus, both nitrogen oxides and SO_2 contribute to the formation of acid rain and PM 2.5. Sulfur dioxide is not implicated in urban ozone or smog.

Sulfur emissions from coal-fired power plants can be controlled by reaction of metal oxide sorbents, such as CaO derived from limestone, with either SO_2 in flue gas or H_2S in producer gas. Hydrogen sulfide can also be removed by reaction with aqueous amine solutions. Biorenewable resources, which typically only contain 0.01 wt-% sulfur, do not require emission controls to meet regulations on SO_2.

Fuel chlorine is converted to hydrogen chloride (HCl) during combustion, which forms hydrochloric acid upon reaction with water droplets. Hydrogen chloride, along with the oxides of nitrogen and sulfur, are known as acid gases. Unlike the other acid gases, HCl has not been implicated in formation of acid rain or PM 2.5. Because it reacts so readily with water, acid will form once flue gas drops below the dew point of the mixture of water and HCl. This usually occurs within the plant where it can corrode equipment. Only a few fuels have sufficiently high chlorine for this to be a problem: refuse-derived fuel because of certain kinds of plastics, and a few agricultural residues, like wheat straw and corn stover. In these cases, scrubbing HCl from the flue gas is important to prevent equipment corrosion.

A more insidious effect arising from fuel chlorine is the formation of dioxin in combustion equipment. Dioxin refers to 210 compounds—75 dioxins and 135 furans—with similar structures and properties. Only 17 of the 210 compounds are toxic, and these differ considerably in their potencies. Some of these are animal carcinogens or have been implicated in birth defects. Combustion-generated dioxin is thought to arise primarily from solid-phase reactions involving metal chlorides and fly ash carbon, reactions that chlorinate aromatic hydrocarbons to produce dioxins. Ironically, high levels of sulfur in flue gas short-circuit the reactions that produce dioxin. For this reason, combustion of refuse-derived fuel, which contains a relatively low ratio of sulfur to chlorine, can generate high dioxin emissions, whereas combustion of coal, which typically has high sulfur-to-chlorine ratios, has not been implicated in dioxin emissions. Examination of Table 9.2 suggests that many herbaceous crops would also generate dioxin unless co-fired with coal. The oxygen-starved conditions of gasification do not produce dioxins.

Incomplete combustion generates *volatile organic compounds (VOC)* and CO. Efforts to reduce NO_x emissions based on staged combustion exacerbate the emission of VOC and CO since they form under fuel-rich conditions. Any fuel, whether derived from fossil resources or biorenewable resources, can emit VOC and CO upon combustion. As previously noted, VOC plays a role in smog formation, while CO contributes to cardiovascular disease. Combustion control, including adjustment of flame position within the boiler and addition of secondary air to complete combustion, usually can achieve satisfactory levels of these products of incomplete combustion.

Heavy metals released from fuels upon combustion typically form oxides or chlorides that exist as particulate matter, although elemental mercury also has been detected as vapor in flue gas. The particulate matter is apportioned between bottom ash, which collects below the grate of certain kinds of boilers, and fly ash, most of which is captured by particulate-control devices, although some escapes into the atmosphere with the flue gas. Coal contains on the order of 0.1 ppm of various heavy metals, while biomass crops grown on uncontaminated soil contain virtually none. Although heavy metals are present in only trace quantities, they are highly toxic and some, like mercury, are *bioaccumulators*; that is, they are taken up by plant or animal tissue and can move through the food chain where they accumulate to toxic concentrations in higher life forms. Proposals to regulate heavy-metal emissions increase the attractiveness of biomass in the production of power.

When burned, solid fuels and heavy fuel oils, whether derived from fossil resources or biorenewable resources, emit primary particulate matter in the form of ash and soot. Uncontrolled emission of these particles into the atmosphere was the most visible manifestation of air pollution before the introduction of particulate control devices. Cyclones, electrostatic precipitators, and bag houses are effective in reducing primary particulate matter. Early regulation focused on particulate matter finer than 10-μm size (PM 10): the most difficult fraction of primary particulate matter to capture within the plant. More recent regulation focuses on PM 2.5, which is chiefly secondary particulate formed outside the plant, since it is considered the more serious health risk.

Properly managed heat and power production results in very little water pollution regardless of the fuel source. Wastewater is generated if fuel is cleaned before it is burned or if spray towers are used to scrub out dust, tar, or other pollutants from producer or flue gas. This wastewater can be a significant source of pollution if not properly handled. Indeed, many of today's federal Superfund Cleanup sites are the legacy of widespread coal gasification at the turn of the century to produce "town gas" for lighting and heating. Tar in the producer gas from these early gasifiers was removed by scrubbing the gas stream with water, which was then disposed of in holding ponds where the tar contaminated the soil. However, modern waste-water treatment technology is able to produce an effluent that does not contaminate the environment.

Production of heat and power is a significant contributor to greenhouse gas accumulation in the atmosphere, including N_2O, CH_4, and CO_2. Although water vapor produced by the combustion of hydrogen-bearing fuels, is also a greenhouse gas, anthropogenic activity is not thought to substantially affect the cycling of water between the Earth's surface and the atmosphere. Nitrous oxide, although a powerful absorber of infrared radiation, can be substantially eliminated by careful combustion control. Methane is not intentionally emitted into the atmosphere, although leaks from natural gas distribution systems can be significant if not carefully managed. The most difficult greenhouse gas to manage in the production of heat and power is CO_2 from the burning of carbon-bearing fuels, including both fossil resources and biorenewable resources. However, since biorenewable resources obtain their carbon by extracting CO_2 from the atmosphere, essentially no net emissions of CO_2 result from growing and burning biorenewable resources. The accounting of net CO_2 emissions is somewhat complicated by the fact that agricultural production currently employs fossil fuels to power tractors and manufacture fertilizer; however, this contribution is small and is expected to diminish as biorenewable resources are increasingly used for the production of transportation fuels and chemicals.

9.4.2 Chemicals and Transportation Fuels

Processing of biorenewable resources to chemicals and transportation fuels, illustrated in Fig. 9.4, can be a significant source of pollution of air, water, and soil, depending on the processes involved and the management of by-products. Because of the diversity of these processes and by-products, it is not possible to provide a comprehensive evaluation of their environmental impacts. Instead, the corn wet-milling process will be used to illustrate the kinds of pollution problems that must be addressed in the production of chemicals and fuels.

Emission of particulate matter is a major problem in grain handling and storage operations at a corn wet-milling plant. These emissions arise from mechanical energy imparted on the corn during conveyance and can be reduced by minimizing grain free-fall distances and grain transport velocities within the plant and by sealing grain receiving areas and process equipment. However, generation of some dust is inevitable, which can require ventilation systems operated in conjunction with cyclones or baghouses to control fugitive dust emissions.

The corn-steep operation in wet-milling plants uses 1.1–2.0 kg of SO_2 per megagram (Mg) of corn. The pungent odor of SO_2 in process water dictates the enclosure and venting of process equipment. Vents can be wet-scrubbed with an alkaline solution to remove SO_2 from the exhaust gas.

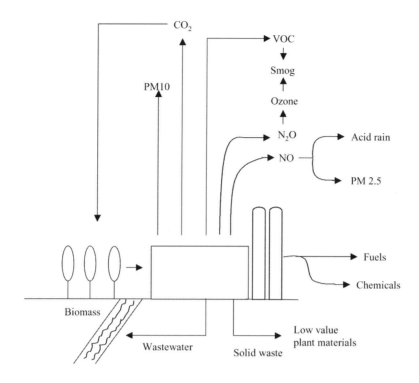

FIG. 9.4. Potential environmental impacts of processing biorenewable resources into chemicals and fuels.

The most objectionable emissions from corn wet-milling plants are VOC released from drying processes. In particular, drying of distillers' dried grains yields VOC with a variety of odors: acetic acid and acetaldehyde produce acrid odors; butyric acid and valeric acid yield rancid odors; and aldehydes produce fruity odors. Moderating drying temperatures can reduce the resulting blue haze from these driers. Further control is achieved with ionizing wet collectors or thermal oxidation of drier vent gases at 750°C for at least 0.5 s.

If the starch product from a corn wet-milling plant is used to produce ethanol as transportation fuel, two other environmental concerns are raised. The first is whether production of ethanol from corn reduces reliance on fossil fuels. Some studies have suggested that more energy is consumed in producing the grain (in the form of nitrogen fertilizer and fuel for agricultural equipment) and converting it to ethanol (especially during the energy-intensive distillation processes) than appears in the final fuel product. At the very least, it is clear that current corn-to-ethanol production and processing systems rely heavily on fossil fuels, a situation that should be addressed if ethanol is to become a major transportation fuel. Closely related to this issue is the net emission of greenhouse gases from the production of ethanol fuel from corn. Although carbon cycles from the atmosphere to the corn crop, then into the ethanol product, and back to the atmosphere upon combustion as transporation fuel, a net positive increase in CO_2 to the atmosphere still results from the fossil fuels used in the production of fertilizers, the operation of tractors and combines, and the distillation of ethanol. To rectify this situation, future biomass-to-ethanol production should reduce the use of fossil fuels.

Modern corn-milling plants produce few water pollution or solid waste disposal problems. Wet corn-milling operations produce wastewater that requires treatment. Aeration ponds are used to oxidize SO2 to sulfate, which then precipitates out into settling ponds. Corn milling produces low-value by-product streams in the form of corn fiber from wet milling and distillers' dried grains and solubles (DDGS) from dry milling. Despite their low value, these by-products find application as feed additives. However, if ethanol production were significantly ramped up to provide transportation fuels, these by-products would be in excess supply and might represent a solid-waste disposal problem. Landfilling of by-product streams is already a problem in the manufacture of high-value chemicals like plant enzymes.

9.4.3 Fibers

The early history of the pulp and paper industry demonstrates that use of biorenewable resources is not inherently environmentally benign. The notorious "paper-town smell" around pulp and paper mills came from reduced sulfur compounds, corrosive and malodorous gases arising from the use of sulfur chemicals in the pulping process. Furthermore, spent-liquor discharges from paper mills turned the waters of local streams black and barren of aquatic life. In the last half of the twentieth century, process improvements have substantially reduced these early problems, although concerns about environmental performance remain, as illustrated in Fig. 9.5.

Of the two major chemical-pulping processes, the sulfite process was originally preferred because of its ability to produce a lighter, easier-to-bleach pulp compared to the sulfate (kraft) process. Increasing environmental regulation gradually shifted the industry toward the kraft process because of its ability to recover chemicals from spent liquor more economically, thus avoiding its discharge into rivers.

However, the kraft process, which requires extensive bleaching to produce white paper, introduced another environmental problem. The most economical bleaching compound is elemental chlorine (Cl_2), which reacts with organic compounds in the spent liquor to form *adsorbable organic halogens (AOX)*. This category includes dioxins and furans, some of which are very toxic to humans and animal life. Many of these organic halogens, discharged to rivers and bays, have found their way into food chains. This problem is being addressed by the adoption of *elemental chlorine-free (ECF) technology*, which uses chlorine dioxide, and *totally chlo-*

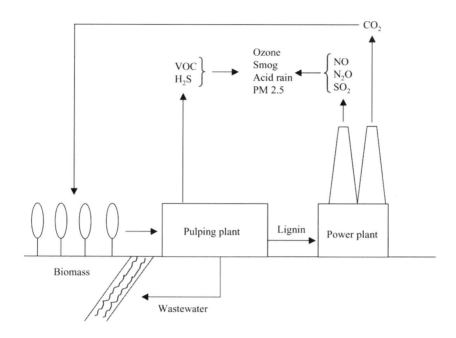

FIG. 9.5. Potential environmental impacts of processing biorenewable resources into fibers.

rine free (TCF) technology, which is based on various combinations of hydrogen peroxide, oxygen, chelating agents, and ozone treatment.

Wastewater contains cellulose fibers, polysaccharides, and various organic and inorganic compounds. Wastewaters are discharged at a rate of 20–250 (m³/t) of *air-dried pulp* (ADP), which is pulp dried to 10 wt-% water. They are high in biochemical oxygen demand (BOD), at 10–40 kg/t of ADP; total suspended solids, at 10–50 kg/t of ADP; chemical oxygen demand (COD), at 20–200 kg/t of ADP; and chlorinated organic compounds, which may include dioxins, furans, and other adsorbable organic halogens, at 0–4 kg/t of ADP.

Primary treatment is performed with a mechanical clarifier or settling pond to recover cellulose fibers, which are then recycled. Chemical precipitation is used to remove certain cations. Secondary treatment is performed in aerated lagoons or anaerobic fermenters to reduce biological oxygen demand (BOD) by over 99%. Tertiary treatment may be performed to reduce toxicity, suspended solids, and color.

The kraft process also contributes to air emission problems. Reduced sulfur compounds, measured as total reduced sulfur, emitted to the atmosphere from the pulping process, include hydrogen sulfide, methyl mercaptan, dimethyl sulfide, and dimethyl disulfide. Volatile organic compounds are emitted during the concentration of black liquor in multiple-effect evaporators. Acid-precipitated lignin from

Table 9.3. Range of pollution emissions (kg/t of ADP*) from pulp and paper mill using the kraft process

Reduced Sulfur	Volatile Organic Compounds	Sulfur Oxides	Nitrogen Oxides	Particulate Matter
0.3–3	15	0.5–30	1–3	75–150

Source: World Bank Group, 1998, "Pulp & Paper Mills," *Pollution Prevention and Abatement Handbook*, World Bank.

* Air-dried pulp.

the concentrated black liquor is used as boiler fuel. The remaining pollution emissions, sulfur oxides, nitrogen oxides, and particulate matter, arise from combustion of the lignin. Table 9.3 lists the range of pollution emissions from pulp and paper mills operating on the kraft process. Emissions are reported as kilograms of pollutant per ton of ADP.

9.5 Utilization

We have seen that the production of biorenewable resources and their processing into biobased products can generate pollutants with adverse impacts on the natural environment. The final products themselves can also have adverse environmental impacts, either in the way they are used or in how they are disposed of at the end of their useful lives. Sometimes these impacts are not obvious until the product has been commercially introduced and its effects on the environment are monitored. Such an approach is not viable if these impacts are potentially large and negative. For this reason, *life-cycle analysis*, which attempts to anticipate the environmental impact of a new product from its manufacture to its final disposal, is increasingly performed before introducing a new product.

Among the most scrutinized biobased products in the United States is fuel ethanol from corn. Its environmental advantages compared to gasoline, including its ability to reduce air pollution and global warming, have been widely touted. However, careful scrutiny reveals that environmental advantages are much harder to justify.

Fuels that contain oxygen, so called oxygenates, are reported to be more "clean-burning" than hydrocarbon fuels since complete combustion is promoted by high oxygen content of the air-fuel mixture. In principle, ethanol, which contains 35 molar-% oxygen, should reduce emissions of CO, particulate matter, and volatile organic compounds. For this reason, the U.S. Clean Air Act Amendments of 1990 mandated the use of gasoline containing oxygenates, called *reformulated gasoline*, to be sold in urban areas that did not meet national targets for reducing urban ozone. The two most commonly employed oxygenates to meet this regulation were *methyl tertiary butyl ether (MTBE)* and ethanol. Methyl tertiary butyl ether has been banned as a fuel-oxygenate in some parts of the world because this toxic compound has become a serious water pollution threat, thus increasing the commercial prospects for ethanol.

Vehicle testing reveals that ethanol is effective in reducing CO emissions from the carbureted engines that were in wide use when ethanol was introduced in the 1970s. However, modern automobiles employing computer-controlled, fuel-injected engines are able to achieve near-complete combustion independent of the oxygen content of the fuel, leading some critics to claim that reformulated gasoline is obsolete. The effect of ethanol on emissions of particulate matter is not well documented.

The actual effect of ethanol on VOC emissions illustrates the importance of life-cycle analysis for new products that potentially have large environmental impacts. The marginal impact of ethanol on CO emissions from modern internal combustion engines might also be expected to be true in the case of VOC emissions, another product of incomplete combustion. In fact, there is evidence that reformulated gasoline based on ethanol exacerbates VOC emissions from automobiles as a result of increased fuel volatility. Although tail-pipe emissions of VOCs might decrease marginally, fuel tank emissions increase significantly: from 34% of total VOC emissions from conventional gasoline to 42% of total VOC emissions from gasoline reformulated with ethanol. Evaporative emissions from fuel tanks are a direct function of fuel vapor pressure. Neat ethanol and high-level ethanol blends with gasoline have lower vapor pressures than gasoline; however, vapor pressure increases as the percentage of ethanol decreases. Reformulated gasoline, which employs only 10% ethanol, has a higher vapor pressure than conventional gasoline thus explaining the higher VOC emissions from the ethanol blends commercially available in the United States. Blending 10% ethanol with 90% gasoline increases the fuel volatility (measured in Reid Vapor Pressure, RVP) over that of conventional gasoline by 3.4–6.9 kPa. A possible solution to this problem is the use of ethyl tertiary butyl ether (ETBE), which does not produce this confounding effect.

Oxygenated fuels may affect NO_x emissions. Although ethanol contains no nitrogen, high oxygen levels promote thermal NO_x formation. On the other hand, the lower flame temperature for combustion of ethanol should favor lower thermal NO_x formation. Engine testing has produced inconsistent results but some environmentalists argue that ethanol-fueled automobiles exacerbate NO_x emissions.

One of the most attractive benefits of using ethanol fuel is reducing global climate change. Since the carbon in ethanol comes from plant material, the argument is that ethanol produces no net CO_2 emissions: an amount of CO_2 equivalent to that released during combustion of ethanol is taken up by crops that will be processed into additional ethanol. This is a potentially important environmental benefit, since conventional transportation fuels account for 27% of anthropogenic CO_2 emissions in the United States. Life-cycle analysis reveals the flaw in this argument: the current system of converting corn starch to ethanol in the United States consumes large quantities of fossil fuels that must be included in the accounting of net CO_2 emissions. Farm machinery is fueled from gasoline produced from petroleum. Nitrogen fertilizer is manufactured in an energy-intensive process from natural gas. Distillation of ethanol requires large quantities of energy obtained from coal or natural gas.

The range of net CO_2 emissions has been estimated at 15–71 kg/GJ, depending on the assumptions made for fuel source and fuel energy requirements for the conversion of corn to ethanol and how by-products are treated. In contrast, gasoline has a net emission rate of about 76 kg/GJ. Thus, some critics claim that the superiority of corn ethanol compared to gasoline in fighting global climate change is marginal. On the other hand, ethanol produced from lignocellulosic materials has lower energy inputs during production, and all process energy comes from feedstock by-products. Net CO_2 emission rates are expected to be only 8–15 kg/GJ from advanced lignocellulose-to-ethanol production and conversion systems.

Further Reading

Plant Science

Ammann, K., Y. Jacot and G. Kjellsson, eds. 2000. *Methods for Risk Assessment of Transgenic Plants: Ecological Risks and Prospects of Transgenic Plants. Where Do We Go from Here? Vol 3.* Boston: Birkhauser.

Letourneau, D.K. and B.E. Vurrows, eds. 2001. *Genetically Engineered Organisms: Assessing Environmental and Human Health Effects.* Boca Raton, Florida: CRC Press.

Lurquin, P.F. 2001. *The Green Phoenix: A History of Genetically Modified Plants.* New York: Columbia University Press.

Production

Hohenstein, W. G., and L. L. Wright. 1994. *Biomass Energy Production in the United States: An Overview, Biomass and Bioenergy* 6: 161–173.

Ranney, J. W., and L. K. Mann. 1994. *Environmental Considerations in Energy Crop Production, Biomass and Bioenergy* 6: 211–228.

Spiro, T. G., and W. M. Stigliani. 1996. *Chemistry of the Environment.* Upper Saddle River, NJ: Prentice Hall.

Processing

Spiro, T. G., and W. M. Stigliani. 1996. *Chemistry of the Environment.* Upper Saddle River, NJ: Prentice Hall.

Wallace, D. 1992. "Grain Handling and Processing." *Air Pollution Engineering Manual.* New York: Van Nostrand Reinhold.

World Bank Group. 1998. "Pulp & Paper Mills." *Pollution Prevention and Abatement Handbook.* World Bank (Available on the World Wide Web at http://wbln0018.worldbank.org/essd/essd.nsf/GlobalView/PPAH/$File/78_pulp.pdf)

Utilization

Ciambrone, D. F. 1997. *Environmental Life Cycle Analysis*, Boca Raton: Lewis Publishers.

Lynd, L. R. 1996. "Overview and Evaluation of Fuel Ethanol from Cellulosic Biomass: Technology, Economics, the Environment, and Policy." *Annu. Rev. Energy Environ.* 21:403–465.

Turnhollow, A. and S. J Kanhouwa. 1993. "Factors Affecting the Market Penetration of Biomass-derived Liquid Transportation Fuels." *Applied Biochemistry and Biotechnology* 39/40:61–70.

Problems

9.1 List four major health and environmental concerns regarding the introduction of transgenic crops to the biosphere.

9.2 Modern agricultural practices have both reduced soil fertility and contributed to global climate change. Explain the relationship between these two environmental impacts.

9.3 Crop production contributes to air pollution not only in the form of exhaust emissions from farm machinery but from dust and gases arising from soil tillage. List the contributions of soil tillage to air pollution.

9.4 Although biomass energy is often touted as environmentally superior to the combustion of coal, this depends on the properties of the biomass and the kind of furnace used to burn it. Provide a qualitative comparison between the air-pollution emissions from a biomass combustor and from a coal combustor.

9.5 Some policy analysts have suggested a tax on CO_2 emissions (the so-called carbon tax) as an incentive for energy consumers to switch from fossil fuels to renewable biomass fuels. This idea is based on the argument that there is no net emission of CO_2 into the atmosphere from biomass, since the same amount of CO_2 is absorbed by biomass in the process of growing it. Assume that coal is available for \$1/GJ while biomass in the form of switchgrass can be grown and delivered to a power plant for \$2.50/GJ. Assume that the coal has a heating value of 28 MJ/kg and contains 70% carbon by weight. How large of tax would have to be placed on CO_2 (\$/ton of CO_2 emitted) before plant operators would consider burning biomass instead of coal?

9.6 The combustion of chlorine-bearing fuels can result in the formation of dioxins, highly toxic pollutants. On the other hand, sulfur in fuel mitigates the formation of dioxin. A rule-of-thumb for preventing dioxin formation is to maintain a sulfur-to-chlorine molar ratio greater than one for fuel mixtures to be burned.

(a) Determine whether dioxin emissions might be expected from the combustion of wheat straw and refuse-derived fuel.
(b) Determine the mass fractions of corn stover and bituminous coal (S/Cl ratio of 5.0) in a fuel blend to avoid dioxin emissions.

9.7 Some environmental activists argue that neither the production of ethanol from cornstarch nor the use of ethanol as an alternative transportation fuel provides environmental advantages.

(a) What is the basis for their argument that the process of converting cornstarch to ethanol is not environmentally sound?
(b) What is the basis for their argument that the utilization of ethanol does not reduce air pollution?

9.8 Describe the major water pollution and air pollution problems associated with modern pulp and paper mills.

Economics of Biorenewable Resources

10.1 Introduction

Market acceptance of new products depends upon the complex interplay of several factors including cost, physical properties, environmental performance, public policy, and cultural prejudices. This chapter focuses on the problem of manufacturing biobased products that are cost-competitive with products already produced from petroleum or other fossil resources. Although biobased products may look attractive from other perspectives, a company will have little incentive to develop them unless the enterprise is projected to be profitable.

Accurate cost forecasting is a difficult and time-consuming activity best left to the experts. However, *cost estimating* is a valuable skill that allows an engineer to obtain "ballpark" approximations of project costs. The goal is to obtain an estimate that is within $+/-$ 30% of the actual cost if the enterprise were pursued. Such estimates are relatively easy to develop.

Two kinds of costs will be considered in this chapter: the cost of producing biorenewable resource feedstocks and the cost of manufacturing biobased products from these feedstocks.

10.2 Estimating the Cost of Feedstock from Biorenewable Resources

For the purpose of cost estimating, biorenewable resources are conveniently classified as either *processing residues* from urban areas, wood mills, or agricultural processing plants; *harvesting residues* from harvesting lumber or agricultural crops; or *dedicated energy crops*. Processing residues are highly concentrated, already having been transported to a central processing facility, and they are often considered to be a waste product. Thus, they can often be acquired at low or even negative cost with minimal transportation cost. Harvesting residues are also under-utilized and represent additional income for producers if they are collected and sold as biorenewable resources. These residues are more expensive than processing residues because they must be collected from fields and transported to a central processing facility. However, development of machinery that simultaneously collects agricultural residues while harvesting the primary crop could reduce costs. *Dedicated energy crops* refers to biorenewable resources grown specifically as feedstock for production of biobased products. Unlike agricultural residues, a dedicated energy crop must bear all the expenses of cultivation and harvest; thus, it is the most expensive of the biorenewable resources. On the other hand, both kinds of residues are very limited in extent, while the cultivation of dedicated energy crops could be expanded significantly, as described in Chapter 3.

The cost of a biorenewable resource is related to the demand for the resource by a supply curve. Figure 10.1 is a generalized representation of a supply curve for the three kinds of biorenewable resources. The least expensive is processing residue, the price of which begins to rise sharply as the supply limit is approached. Notice that the supply is relatively smaller than the other kinds of biorenewable resources. Because of additional collecting and harvesting expenses, the cost of harvesting residue is significantly higher than for processing residue. The supply, however, is substantially greater. Finally, the costs of a dedicated energy crop are higher than the other kinds of biorenewable resources but the cost climbs more gradually as the demand increases. The increasing price reflects the use of less productive land to supply the additional demand. Notice that the supply curve extends much further than the other supply curves.

The cost of biorenewable resources is highly variable and dependent on local conditions of supply and demand. This is particularly true for the processing residues and wastes, and no effort will be made here to develop a methodology for

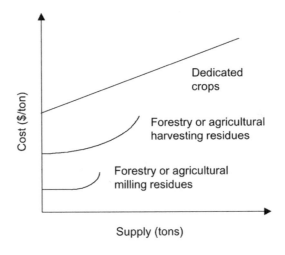

FIG. 10.1. Example of supply curves for different kinds of biorenewable resources.

Table 10.1. Availability and cost of potential feedstocks

Feedstock	Production (10^6 tons/yr)	Price (1994 \$/kg)
Corn	191	0.09
Potato	17	0.16
Sorghum	16	0.09
Beet molasses	1	0.09
Cane molasses	1	0.09
Sugar cane	25	0.03
Agricultural residues		
Low cost	4	12.9
Mid cost	36	38.8
High cost	50	47.4
Forest residues–logging		
Low cost	3	12.9
Mid cost	3	25.9
High cost	3	43.1
Forest residues–mill	3	17.2
Municipal solid waste		
Mixed paper	26.0	0–19
Packaging	14	0–5.2
Urban wood	3.5	12.9–25.9
Yard waste	11	0–12.9

Sources: Crop data from K. Polman, 1994, "Review and Analysis of Renewable Feedstocks for the Production of Commodity Chemicals," *Appl. Biochem. & Biotech.* 45/46:709–722; Waste and residue data from L. R. Lynd, 1996, "Overview and Evaluation of Fuel Ethanol from Cellulosic Biomass: Technology, Economics, the Environment, and Policy," *Annual Rev. Energy Environ.* 21:403–465.

estimating their costs. Instead, Table 10.1 is included to provide an estimate of the availability and cost of several kinds of residues and wastes along with a comparison of the costs of a few crops.

The cultivation and harvesting of dedicated energy crops, on the other hand, is amenable to standardized cost estimating since information on "unit operations," such as planting, fertilizing, and harvesting, can be readily obtained from knowledgeable sources.

10.2.1 Unit Cost for Production of Annual Crops

This methodology focuses on *annual crops* such as corn or sweet sorghum, which are planted and harvested every year. The methodology can be easily adapted to estimating the cost of harvesting residues. The methodology consists of breaking down a production system into important expense categories and assigning a cost per hectare. As shown in Table 10.2, these expense categories include preharvest machinery, seed/cuttings, fertilizer, pesticides, crop insurance, interest on short-term loans, miscellaneous, harvest machinery, labor, land, and transportation. Expenses for each expense category allow for both variable costs and fixed costs.

Table 10.2. Cost of production for annual crop

Crop:			Yield: (Mg/ha/yr)		Date:
Production Method:			Market Price: ($/Mg)		
			Expenses		Explanation
	$/Unit	Unit/ha	Variable ($/ha)	Fixed ($/ha)	
Preharvest Machinery					Plowing, disking, fertilizing, planting, cultivating, and spraying. Variable expense is fuel and repair. Fixed expense is capital charges.
Seed/Cuttings					
Fertilizer					
Nitrogen					
P_2O_5					
K_2O					
Lime					
Pesticides					
Herbicides					
Insecticides					
Crop Insurance					
Interest					Interest on preharvest variable expenses
Miscellaneous					
Harvest Machinery					Variable expense is fuel and repair. Fixed expense is capital charges.
Harvest					May involve combining or mowing, raking, and baling or forage chopping depending on the type of crop
Haul					
Dry					
Labor					Assumed to be hired labor, a variable expense
Land					A fixed expense whether rented or financed
Transportation					Set to zero when calculating "farm-gate" price of crop
Total fixed or variable expense					Add variable and fixed expenses
Total expense					
Unit Production Cost ($/Mg)					Divide total costs ($/ha) by yield (Mg/ha)

Variable costs depend on the extent of their usage. Things such as seed/cuttings, fertilizer, and pesticides are *variable costs*. There are also *fixed costs*, which are invariant during the operation and include such things as land rental and taxes. Some cost categories have both fixed and variable costs, of which machinery expenses are the most prominent example. Variable costs of machinery include fuel for operating machinery and repair to machinery. The fixed cost of machinery is primarily payment of interest and principal on loans used to purchase the machinery.

Preharvest machinery expenses include plowing, disking, planting, fertilizing, cultivating, and spraying. *Harvest-machinery expenses* vary with the kind of crop being harvested. Raising grain crops involves combining; hay crops require a series of operations, including cutting, raking, and baling; forage crops are harvested with a forage chopper; and short-rotation woody crops require specialized cutting and chipping machinery. Expenses associated with several kinds of production operations on a hectare basis can be estimated from Table 10.3, which covers variable and fixed expenses of machinery operation in the state of Iowa. These numbers do not include labor costs, which are estimated separately. More detailed information can be obtained from extension services of many land-grant universities.

The *cost of seed and cuttings* is calculated as the product of cost per unit of seed or cutting and units planted per hectare. The unit of seed or cutting is either the number of kernels, in the case of corn, the weight of seed, or the number of cuttings, in the case of short-rotation woody crops. The units planted per hectare for a particular crop depend on climate and soil type. Both cost per unit and units planted per hectare can be estimated for several kinds of crops from Table 10.4. The *cost of fertilizer and pesticides* per hectare can be estimated from Table 10.5 for several kinds of crops. More detailed information for specific agricultural regions can be obtained from extension services of many land-grant universities.

Crop insurance protects a producer against lost income in case of catastrophic crop loss associated with damage from hail, wind, or flooding. *Interest* includes the cost of money borrowed for purchase of seed and chemicals and other preharvest variable expenses. These are short-term loans for a period roughly equal to the time between planting and harvesting an annual crop. Financing of perennial crops requiring several growing seasons before harvest requires longer-term financing and more sophisticated cost analysis, as described later. *Miscellaneous expenses* may include property taxes or other expenses not accounted for in the other expense categories.

Labor rates are determined by adding up the time required to perform all preharvest and harvest operations and multiplying by the hourly wage of laborers. For many cropping systems, the total labor requirement has already been determined; Table 10.6 provides a partial tabulation of such information. Hourly wages may vary considerably depending on labor availability and the skill required for the operation.

The cost of renting land or financing the purchase of land can account for more than half the cost of production. In the case of financed purchase of land, this expense is called a *capital charge* and represents annual payments of principal and interest on a long-term loan. For a loan on the *principal* amount, P, taken out at

Table 10.3. Machinery costs for crop production

	Expenses ($/ha)*	
	Fixed	Variable
Moldboard plow	17.67	9.91
Chisel plow	7.98	3.78
Chop stalks	10.75	5.56
Tandem disk	6.18	2.64
Offset disk	11.17	4.60
Peg tooth harrow	2.47	0.99
Sprayer/disk	6.80	2.94
Field cultivator	6.45	1.75
Bulk fertilizer spreader	4.94	1.63
NH_3 applicator	6.65	3.31
Chisel plow, NH_3 application	9.59	6.08
Grain drill	17.07	3.81
Broadcast seeder	7.68	3.63
Planter	11.59	3.51
No-till planter	13.91	4.32
No-till drill	22.93	7.26
Rotary hoe	2.30	0.99
Cultivator	4.55	2.30
Sprayer	2.72	1.31
Combine corn	30.81	20.66
Combine beans	28.79	13.42
Combine small grain	32.10	14.16
Haul grain (on farm)	0.02/bu	0.01/bu
Corn picker	39.91	22.19
Silage harvester	46.33	29.78
Haul silage	0.75/Mg	0.44/Mg
Mower-sickle	9.88	4.74
Mower-conditioner	10.67	5.58
Rake	5.98	3.58
Square baler (including twine)	0.17/bale	0.13/bale
Round baler (including twine)	2.41/bale	2.74/bale
Stacker	8.87	6.99
Windrower	9.69	4.94
Haul hay and straw	0.87/Mg	0.52/Mg
Forage chopper	27.77	18.09
Forage blower	0.56/Mg	0.19/Mg

Source: Adapted from M. Duffy, *Estimated Costs of Crop Production in Iowa–2002* (Iowa State University Extension Publication 1712).

* Units are $/ha unless otherwise noted.

interest rate, i, to be paid back over a period of n years, the *annual capital charge*, A_{CC}, that would appear in this expense category is:

$$A_{CC} = Pi(1 + i)^n/[(1 + i)^n - 1] \qquad (10.1)$$

Unit cost of production is calculated as the total expenses ($/ha) divided by the crop yield (Mg/ha). An important expense neglected in this analysis is federal

Table 10.4. Costs for seed and cuttings in crop production

Seed/cutting	Unit	Unit Cost ($/unit)	Application (units/ha)
Corn	kernals (k)	0.001	22,000
Soybeans	kilogram	0.66	67.2
Soybeans (GMO)	kilogram	0.93	67.2
Oats	kilogram	11.03	2.2
Alfalfa	kilogram	6.62	9.0
Bromegrass	kilogram	2.21	6.7
Orchardgrass	kilogram	2.76	3.4
Short-rotation woody biomass	cutting	0.1	7200

Source: Data for agricultural cropping systems adapted from M. Duffy, *Estimated Costs of Crop Production in Iowa—2002*, Iowa State University Extension Publication 1712; data for forestry cropping systems adapted from G. Wiltsee and E. Hughes, October 1995, "Biomass Energy: Cost of Crops and Power," *Electric Power Research Institute Final Report TR-102107–* Vol. 2.

Table 10.5. Costs for chemicals in crop production

Fertilizer	Unit	Unit cost ($/unit)	Application (units/ha)				
			Corn	Soybeans	Alfalfa	Cool-season grass	SRWC*
Nitrogen	kg	0.46	135	0	0	90	137
Phosphate	kg	0.55	45	34	39	34	see note
Potash	kg	0.29	34	67	140	0	see note
Lime	Mg	14.30	1.12	1.12	1.12	0	see note

Pesticide	Unit Cost ($/ha)				
	Corn	Soybeans	Alfalfa	Cool-season grasses	SRWC*
Herbicide	77	77	28	11	85
Insecticide	35	0	0	0	see note

Source: Data for agricultural cropping systems adapted from M. Duffy, *Estimated Costs of Crop Production in Iowa–2002*, Iowa State University Extension Publication 1712; data for forestry cropping systems adapted from G. Wiltsee and E. Hughes, October 1995, "Biomass Energy: Cost of Crops and Power," *Electric Power Research Institute Final Report TR-102107– Vol. 2*.

*Short-rotation woody crop—nitrogen applied every third year, other fertilizers dependent on soils; herbicide is applied during establishment with occasional follow-up at $38/ha; insecticide applied as needed at $2.15/ha.

income tax that must be paid on the profits resulting from the enterprise. However, this tax is often ignored in calculating unit cost of production since it depends on all income streams for a tax-paying individual or corporation and is complicated by various depreciation allowances and accounting methodologies. Income tax will be accounted for in the cash flow analysis described later for estimating unit cost of production for perennial crops.

Table 10.7 provides a cost projection for corn grown in Iowa in 2002. This analysis assumes conventional tillage following a crop of corn in the previous year,

Table 10.6. Labor rates for crop production

Crop/Cropping System	Labor (hours/ha)
Corn following corn	7.0
Corn following soybeans	6.4
Corn following soybeans (no-till)	5.7
Corn silage following corn	12.4
Soybeans following corn	6.1
Soybeans following corn (no-till)	4.3
Genetically modified soybeans (till)	5.6
Genetically modified soybeans (no-till)	4.3
Alfalfa seeded with oat companion crop (establishment year)	2.5
Alfalfa seeded with oat companion crop (after establishment)	7.4
Perennial grass seeded with herbicide (establishment year)	2.5
Harvesting perennial grass as large round bales (per cutting)	3.3
Harvesting perennial grass as small square bales (per cutting)	4.4
Establishing short-rotation woody crops	1.2
Maintenance of short-rotation woody crops (per year)	1.2
Harvesting short-rotation woody crops	7.4

Source: Data for agricultural cropping systems adapted from M. Duffy, *Estimated Costs of Crop Production in Iowa–2002*, Iowa State University Extension Publication 1712; data for forestry cropping systems adapted from G. Wiltsee and E. Hughes, October 1995, "Biomass Energy: Cost of Crops and Power," *Electric Power Research Institute Final Report TR-102107–Vol. 2.*

which affects the amount of field preparation required. Notice that the largest variable expenses are nitrogen fertilizer ($62.10/ha) and herbicide ($76.60/ha). However, rental of land, a fixed expense, is the most important factor determining the cost of production, representing one-third the cost of producing corn. Also notice in this example that the cost of production at $128.52/Mg is 50% higher than the market price for corn in early 2002. This example highlights some of the difficulties facing U.S. producers in the early twenty-first century as a result of overproduction relative to market demand and competition from producers in developing nations who have lower fixed costs.

10.2.2 Unit Cost for Production of Perennial Crops

Perennial crops, such as hybrid poplar and switchgrass, have planting/harvest cycles that span several years and may show dramatic differences in production expenses and revenues from one year to the next. In this case, calculating production costs are more complicated because capital investment is required early in the project, while significant revenue may not be generated for two or three years, in the case of switchgrass, or as long as seven years, in the case of dedicated woody crops. Furthermore, meaningful analysis requires an accounting of the time value of money, which recognizes that a dollar spent or earned today is worth more than the same dollar in the future as a result of inflation and investment opportunities for money in the present.

Table 10.7. Example of cost of production for corn crop

Crop: Corn Following Corn			Yield (Mg/ha/yr): 6.3*	Date: 2002
Production Method: Chisel plow, tandem disk, apply N, field cultivate, plant, cultivate, and spray			Market price ($/Mg): $82.50**	
			Expenses	
			Variable ($/ha)	Fixed ($/ha)
Preharvest Machinery			18.61	46.11
	$/Unit	Unit/ha		
Seed	$1/1000 k	54,400	54.40	
Fertilizer				
Nitrogen	$0.46/kg	135	62.10	
P_2O_5	$0.55/kg	45	24.75	
K_2O	$0.29/kg	34	9.86	
Lime	$14.30/Mg	1.2	17.16	
Pesticides				
Herbicides			76.60	
Insecticides			34.59	
Crop Insurance			14.83	
Interest			16.26	
Miscellaneous			14.83	
Harvest Machinery				
Harvest			20.66	30.81
Haul			2.47	4.94
Dry			35.01	9.88
Labor			56.34	
Land			—	259.46
Transportation			—	—
Total fixed or variable expense			458.47	351.20
Total expense			809.67	
Unit Production Cost ($/Mg)			128.52	

Source: M. Duffy, *Estimated Costs of Crop Production in Iowa - 2002*, Iowa State University Extension Publication 1712.

*Equivalent to 100 bushels per acre.

**Equivalent to $2.10 per bushel.

The relationship between the *future value*, FV, and *present value*, PV, of a sum of money compounded annually at an interest rate, i, for a total of n years is:

$$FV = PV(1 + i)^n \tag{10.2}$$

For example, $100 invested today at 5% for 2 years will return:

$$FV = \$100 \times (1 + 0.05)^2 = \$110 \qquad (10.3)$$

Similarly, inflation at 5% per annum makes $100 worth of merchandise today cost $110 in two years.

Discounting is the opposite of compounding: the present value of a sum of money to be spent or received in the future is reduced according to:

$$PV = FV/(1 + i)^n \qquad (10.4)$$

where i is now referred to as the *discount rate*. Thus, if the discount rate is 5% per annum, $100 to be spent two years hence has a present value calculated as follows:

$$PV = \$100/(1 + 0.05)^2 = \$91 \qquad (10.5)$$

Since the present value of money to be spent in the future declines with increasing interest rate and number of years into the future, the importance of accounting for the time at which money is spent or received is evident.

Cash flow analysis is a method for accounting for the time value of money. The procedure is quite simple. For each year, all inflows of cash are subtracted from all outflows of cash to obtain an annual cash flow. In the case of crop production, inflows and outflows are conveniently expressed on a per hectare basis.

The production cost procedure developed for annual crops has been modified in Table 10.8 to allow an accounting of production costs of perennial crops over several years using the cash-flow method. Notice that the table still includes the major expense categories previously considered, but columns have been added for fixed and variable expenses for every year of the enterprise (assumed to be five in Table 10.8, but possibly even longer for some short-rotation woody crops). Furthermore, additional rows are included for calculating cash flow, discounted cash flow, and net present value.

Inflows consist of *annual revenues*, A_R, from the sale of products or co-products. These revenues are calculated from market price and yield of the crop:

$$A_R(\$/ha) = \text{Market price } (\$/Mg) \times \text{Yield } (Mg/ha) \qquad (10.6)$$

Prices vary considerably from one year to the next, and yields are strongly dependent on soil type, geographic location, and weather. Therefore, care should be taken in assigning values for the purpose of cost estimating.

Outflows are *operating expenses*, A_{OE}, owner's *capital invested* in the purchase of equipment or property, A_{CI}, and *federal income tax*, A_{IT}. These expenses are subtracted from revenues to yield *net cash flow*, A_{CF}:

Table 10.8. Cost of production for perennial crop

Crop:										Yield (Mg/ha/yr):	Date:
Production Method:										Market Price ($/Mg):	
	Expenses ($/ha)										
Year:	1		2		3		4		5		
	Fixed	Variable	Fixed	Variable	Fixed	Variable	Fixed	Variable	Fixed	Variable	
Capital Investment, A_{CI}											Owner's capital to purchase land or equipment
Preharvest Machinery											Fixed expense includes interest A_I and principal A_P on loans
Seeds/cuttings											
Fertilizer											
Pesticides											
Crop Insurance											
Interest											
Miscellaneous											
Harvest Machinery											Fixed expense includes interest A_I and principal A_P on loans
Labor											
Land											If land is financed, includes interest A_I and principal A_P on loans
Total Fixed, A_{FE} or Variable, A_{VE} Expenses											
Total Operating Expenses, A_{OE}											$= A_{FE} + A_{VE}$
Total Revenue, A_R											$=$ Market price $\times$ yield
Federal Income Tax, A_{TT}											$=$ Tax rate $\times$ $(A_R - (A_{OE} - A_P) - A_D)$
Cash Flow, A_{CF}											$= A_R - (A_{OE} + A_{CI} + A_{TT})$
Discounted Cash Flow, A_{DCF}											$= A_{CF}/(1 + r)^n$
Net Present Value, NPV											$= \Sigma\ A_{DCF}$

$$A_{CF} = A_R - (A_{OE} + A_{CI} + A_{IT}) \tag{10.7}$$

Notice that capital invested is money provided by owners of the enterprise, whereas capital charges, an expense accounted for under operating expenses, are the annual interest and principal payments the owners make toward retiring a loan on the purchase of land and equipment.

Federal income taxes are assessed as a certain percentage of the difference between revenues, A_R, and certain allowable expenses and deductions. *Allowable expenses* include all the previously detailed operating expenses except for repayment of principal on loans, A_P.

Allowable deductions are depreciation valuations on capital equipment. *Depreciation* is an important tax incentive based on the idea that equipment purchased for an enterprise "wears out" over time. Depreciation allows investors to exclude from taxation the loss in value of their original capital investment. Although different rules for calculating depreciation have been devised, straight-line depreciation is commonly employed. The annual amount of depreciation, A_D, is calculated by dividing the difference in *fixed capital cost*, C_{FC}, and *salvage value*, S, of the equipment by the *depreciating time period*, t_D. The depreciation period is usually taken as the useful life of the equipment or as otherwise allowed by tax laws. (Real estate and working capital, consisting of start-up inventory for a plant, cannot be depreciated.):

$$A_D = (C_{FC} - S)/t_D \tag{10.8}$$

The calculation of income taxes is made by the following relationship:

$$A_{IT} = \text{tax rate} \times (A_R - (A_{OE} - A_P) - A_D) \tag{10.9}$$

Tax rates are strongly dependent on the income generated by the enterprise. Small agricultural enterprises may only be taxed at 15–25% whereas large enterprises have been historically taxed at 50%.

Cash flows must be discounted for the year in which cash is spent or earned. *Annual discounted cash flow*, A_{DCF}, is calculated for each year n:

$$A_{DCF} = A_{CF}/(1 + i)^n \tag{10.10}$$

The *discount interest rate* to use in this equation is one appropriate to the investment goals of the owners. It normally accounts for both the *inflation rate*, f, of money for the period of investment as well as a *real rate of return*, r. This so-called *nominal rate of return*, i, is obtained as the geometric mean of the inflation rate and real rate of return:

$$1 + i = (1 + f) \times (1 + r) \tag{10.11}$$

An excellent approximation to this relationship is:

$$i = f + r \tag{10.12}$$

Summing over all the annual discounted cash flows yields the *net present value (NPV)* of the enterprise:

$$NPV = \sum_n A_{DCF} \tag{10.13}$$

Clearly, a positive net present value at the end of the investment period indicates an investment that exceeds the desired discount rate while a negative value indicates one that does not achieve the desired rate. If the discount rate is adjusted to achieve a NPV of exactly zero, the resulting interest rate is called the *discounted cash flow rate of return (DCFRR)*. Alternatively, if the market price is adjusted to achieve a NPV of exactly zero, the resulting market price represents the production cost of the crop assuming the specified discount rate. This second approach is particularly useful in estimating the cost of feedstocks from biorenewable resources grown as perennial crops. It can also be employed to calculate the cost of producing biobased products over the life of a manufacturing facility.

A simplified example of cash flow analysis for a perennial crop is presented in Table 10.9. The crop is assumed to be a perennial grass that is planted in the first year and harvested annually in the second through fifth years of the enterprise. Capital invested in the first year, for the purchase of land, is $700/ha (dollars enclosed in parentheses represent negative cash flows). The discount rate is chosen to be 10%, which represents the expected rate of return of investors in the enterprise. Annual expenses, which are not detailed in this example, are assumed higher in the first year than subsequent years because of the cost of planting and chemical treatment (herbicide and fertilizer). Income taxes are ignored to simplify the analysis. Revenue, which starts in the second year, of $540/ha is based on a constant market price of $45/ton and crop yield of 12 ton/ha (obviously, both price and yield will vary from year to year in a real enterprise).

Table 10.9. Simplified example of cash flow analysis for perennial crop production

Crop: Perennial Grass	Yield (tons/ha): 12	Market Price($/ton): 45				
Discount Rate: 10%						
Year:		1	2	3	4	5
Capital Invested	A_{CI}	($700)	$0	$0	$0	$0
Expenses	A_E	($300)	($250)	($250)	($250)	($250)
Revenue	A_R = Yield × Price	$0	$540	$540	$540	$540
Annual Cash Flow	$A_{CF} = A_R - A_{CI} - A_E$	($1,000)	$290	$290	$290	$290
Discounted Annual Cash Flow	$A_{DCF} = A_{CF}/(1+i)^n$	($909)	$240	$218	$198	$180
Net Present Value	$NPV = \Sigma A_{DCF}$					($73)

Annual cash flow in the first year is negative because no crop is harvested during the establishment year. Cumulative revenue in subsequent years must be high enough to not only off-set the expenses of those years but to pay for the annual expense and capital investment of the first year. To get a realistic appraisal of the time value of money, the discounting factor of $(1 + i)^{-n}$ is applied to the cash flow of the nth year to derive discounted cash flows for each year. Table 10.9 shows that revenue in early years is more valuable than revenue in later years while expenses in early years are more costly than expenses in later years.

The cash flows over the five years are summed to yield the net present value of the enterprise after five years, which in this example is $-\$73/ha$. Thus, the enterprise has not returned the investors their expected return of 10% per annum. In this simple example, we can see that one additional year of harvesting would yield a net present value that was positive, unless, of course, the perennial grass had to be replanted after five years.

Production cost of this perennial grass over five years is found by plotting net present value of the enterprise after five years vs. assumed market price of the crop, as shown in Fig. 10.2. When the market price reaches approximately \$47, net present value becomes zero, representing the cost of producing the crop in the case of a 10% annual rate of return.

Table 10.10 gives estimated costs (1990 dollars) of producing several herbaceous and woody crops using discounted cash-flow analysis assuming a real after-tax return of 12%. The woody crops (sycamore and poplar) assumed six-year rotations while other species were treated as herbaceous crops. Research and genetic improvements

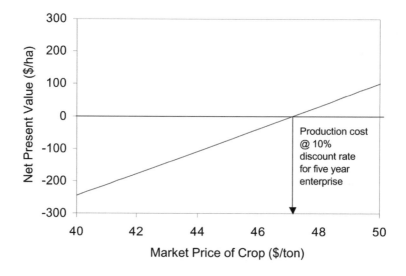

Fig. 10.2. Estimating unit cost of production for a perennial crop using cash-flow analysis (based on data in Table 10.9).

Table 10.10. Estimated U.S. delivered costs for candidate biomass energy crops in 1990 and 2030

Region and Species	1990		2030	
	Yield (Mg/ha-yr)	Cost ($/Mg)	Yield (Mg/ha-yr)	Cost ($/Mg)
Great Lakes				
Switchgrass	7.6	104.07	15.5	61.32
Energy sorghum	15.5	62.56	30.9	36.79
Hybrid poplar	10.1	113.79	15.9	72.82
Southeast				
Switchgrass	7.6	105.89	17.3	52.1
Napier grass	13.9	63.72	30.9	33.31
Sycamore	8.1	88.631	14.3	53.19
Great Plains				
Switchgrass	5.4	74.32	10.3	44.05
Energy sorghum	6.3	91.73	13.7	48.07
Northeast				
Hybrid poplar	8.1	105.26	11.9	71.69
Pacific Northwest				
Hybrid poplar	15.5	66.69	23.8	44.73

Source: Adapted from M. D. Fraser, 1993, in "Energy from Biomass and Wastes XVI," D. L. Klass, ed. (Chicago: Institute of Gas Technology), 295.

are expected to increase yields by 2030. Notice that production costs in 1990 for these non-traditional crops are 10–50% less than production costs for corn using traditional production practices (see Table 10.7), while herbaceous and woody crops grown in 2030 are expected to be 75% cheaper than corn as a result of research and genetic improvements. These non-traditional crops may prove attractive to both producers and manufacturers if they can be processed into high-value biobased products.

Cash flow analysis is an essential tool for estimating unit cost of production for perennial crops. It also is a powerful technique for estimating costs of annual crops, even for relatively constant revenues and expenses, if the time value of loans or invested capital is to be accounted for over the life of the loan or investment period. Cash flow analysis is also very important in estimating costs of manufacturing biobased products from biorenewable resources.

10.3 Estimating Unit Cost for Manufacturing Biobased Products

There are two important quantities to determine in the economic analysis of a process: capital cost and operating cost. *Capital cost* is the amount of money to build a plant or facility and includes all equipment and labor associated with installation of the equipment. *Operating cost* represents the annual expenses to keep a plant or facility in full production. It includes costs of feedstock and fuel, labor to operate the plant, and payment of principal and interest on loans (capital charges).

Operating cost is often expressed as *production cost*, which is annual operating cost divided by annual production output.

The first step, of course, in cost estimation is to design the process to be employed in the plant or facility, whether a biomass gasification power plant or a fermentation facility to produce lactic acid. The design should include a flow sheet that quantifies temperatures, pressures, and balances of mass and energy through the process. This information is key in specifying the cost of equipment to be purchased and installed and raw materials and utilities to be consumed in the operation of the plant. After this is done, capital costs and operating costs can be estimated.

10.3.1 Capital Costs

Capital costs are broken down into four major categories: direct project expenses, indirect project expenses, contingency and fee, and auxiliary facilities. Each of these major categories can be broken down into subcategories, as illustrated in Table 10.11, to yield a summary of capital costs.

10.3.1.1 Direct Costs

Direct costs, or direct project expenses, include the purchase price of equipment to be installed, cost of materials required for the installation and wages for installation labor.

Equipment

The cost of equipment is usually expressed in terms of *free on-board (f.o.b.)*, which is the price paid to a supplier to crate and place equipment on-board a freight carrier. Freight expenses are accounted for under indirect project expenses. The f.o.b. equipment cost is given the symbol, C_p, and is used to estimate other capital costs. These other costs can be 3–5 times more than the cost of the equipment.

Equipment costs can be determined by calling various suppliers and obtaining informal or formal quotes. Most will supply catalogs upon request and many maintain Web sites, which may or may not include price information. A good way to identify suppliers is the Thomas Register, the "Yellow Pages" of industry, which is available in most technical libraries and on the Internet. Equipment cost estimation for new technologies or complex projects can be done more quickly with tabulations found in the literature [1–5]. These are usually presented as charts of *purchased equipment cost (f.o.b)* vs. an appropriate sizing parameter (volume, heat transfer area, flow rate, power, etc.).

Very frequently cost data are not readily available for the particular size of equipment or facility of interest, and a method for scaling the cost to the appropriate size is required. However, assuming that costs of equipment are linearly related to the size of the equipment will almost always produce large errors. The *principle of economies of scale* predicts that capital costs escalate proportionally slower than the size of the facility; hence, the unit cost of a product generally decreases as the facility becomes larger.

Table 10.11. Summary of capital costs

	Cost	Calculation	Description
Direct project expenses			
Equipment (f.o.b.)		C_P	"Free on board"—cost when placed aboard the shipping carrier. Obtained from suppliers or estimated from tabulations of purchased equipment costs or installed bare module costs (for example, Ref. 3). Adjusted for size and inflation.
Materials for installation		$C_M = MMF \times C_P$	Based on tabulated materials module factors (MMF) found in References 1 and 2; expressed as fraction of C_P
Direct labor		$C_L = LMF \times (C_P + C_M)$	Based on tabulated labor module factors (LMF) found in References 1 and 2; expressed as fraction of $C_P + C_M$
Total direct		$C_D = C_P + C_M + C_L$	
Indirect project expenses			
Freight, insurance, taxes		$C_{FIT} = 0.08C_P$	Estimated as 8% of total purchased equipment costs
Construction overhead		$C_O = 0.7C_L$	Estimated as 70% of total labor installation costs
Engineering expenses		$C_E = 0.15(C_P + C_M)$	Estimated as 15% of total equipment and materials costs
Total indirect		$C_{ID} = C_{FIT} + C_O + C_E$	
Bare module cost		$C_{BM} = C_D + C_{ID}$	Defined as combined direct and indirect costs
Contingency & fee		$C_{CF} = 0.18C_{BM}$	Estimated at 18% of bare module cost
Total module cost		$C_{TM} = C_{BM} + C_{CF}$	Defined as bare module cost plus contingency and fee
Auxiliary facilities		$C_{AF} = 0.3C_{TM}$	Estimated as 30% of total module cost (when auxiliary facilities are required)
Grassroots capital		$C_{GR} = C_{TM} + C_{AF}$	Defined as total module cost plus auxiliary facilities cost

A rational basis of comparison must scale capital costs to realistic equipment size. Recognizing that the capacity of a simple piece of equipment, like a water tank, increases with volume (that is, with the cube of the characteristic dimension), and that the cost of the tank for a given wall thickness increases with the surface area of metal plate used in its construction (that is, as the square of the characteristic dimension), it then follows that the cost of the tank increases as the 2/3 power of capacity. This idea is expressed as the following simple scaling law [3]:

$$C_{p,s} = C_{p,b}(S_s/S_b)^n \qquad\qquad (10.14)$$

where $C_{p,s}$ = predicted cost of the specified equipment
$C_{p,b}$ = known cost of the baseline equipment
S_s = size of the specified equipment
S_b = size of the baseline equipment
n = economy of scale sizing exponent (less than unity)

The *economy of scale sizing exponent* diverges from the theoretical 2/3 value depending on the kind of equipment and sometimes on the size range of the equipment. This factor is reasonably well known from industrial practice for a variety of parts and equipment; these can be used to estimate overall costs of systems made up of such parts and equipment. Table 10.12 includes unit costs for several kinds of process equipment, adjusted to 2002 dollars, along with corresponding sizing exponents for scaling the equipment to other sizes. Keep in mind that special materials of construction, such as stainless steel, or operation at elevated pressures or other special circumstances may substantially increase equipment costs compared to the values presented in Table 10.12. In these cases more detailed tables [1–3] should be consulted to obtain more accurate cost estimates.

> **Example:** A boiler is to be purchased to generate low-pressure steam (100 kPa). What is the f.o.b cost for a boiler with 1×10^6 kg/hr steam capacity? What would be the cost of a replacement pump and driver for this boiler?
>
> From Table 10.12 a unit cost of $4557 for a low-pressure boiler with 1 kg/hr of steam flow is read. Linear scaling would suggest the desired steam capacity would cost:
>
> $$\frac{\$4557}{kg/hr}(1 \times 10^6 \text{ kg/hr}) = \$4.6 \text{ billion dollars}$$
>
> Fortunately, economies of scale result in a considerably smaller escalation of price with increasing equipment size. Also from Table 10.12 a sizing exponent of 0.5 is obtained. Substituting this information into equation 10.14 yields:

Table 10.12. Unit costs, scaling exponents, materials module factors (MMF), and labor module factors (LMF) for various kinds of plant equipment

Device	Sizing Parameter	Unit	Unit Cost* ($)	Sizing Exponent	MMF	LMF
Process furnaces	Heating rate	kW	810	0.85	0.34	0.22
Direct-fired heaters	Heating rate	kW	73	0.85	0.31	0.22
Shell & tube heat exchangers	Heat transfer area	m^2	2,400	0.65	0.72	0.37
Process vessel (vertical)	Volume	m^3	6,000	0.71	1.04	0.49
Process vessel (horizontal)	Volume	m^3	6,200	0.6	0.65	0.39
Pump & driver	Flow rate × pressure head	m^3 kPa/min	350	0.52	0.71	0.42
Compressor & driver	Power	kW	5,300	0.75	0.60	0.39
Agitators (propeller)	Power	kW	2,100	0.5	0.28	0.27
Air dryers	Volumetric flow rate	m^3/min	7,600	0.56	0.27	0.37
Blowers & fans	Volumetric flow rate	m^3/min	79	0.68	0.27	0.25
Blenders	Volumetric flow rate	m^3/min	28,000	0.52	0.27	0.27
Boilers (100 kPa)	Mass flow rate steam	kg/h	3,100	0.5	0.19	0.26
Boilers (4000 kPa)	Mass flow rate steam	kg/h	4,300	0.5	0.19	0.26
Centrifuges	Diameter	m	63,000	1	0.28	0.23
Conveyer belt (0.6 m width)	Length	m	6,000	0.65	0.27	0.33
Conveyer bucket (30 tph)	Length	m	2,500	0.65	0.28	0.44
Conveyer screw (0.3 m dia.)	Length	m	3,600	0.8	0.27	0.25
Crushers (pulverizer)	Mass flow rate	kg/h	3,500	0.35	0.27	0.25
Crystallizers (forced circulation)	Mass flow rate	tpd	43,000	0.55	0.27	0.38
Dryers (rotary, direct)	Volume	m^3	16,000	0.42	0.28	0.36
Dryers (rotary, vacuum)	Volume	m^3	36,000	0.69	0.28	0.36
Duct work, shop-fabricated, aluminum	Length	m	54	0.55	0.00	0.87
Duct work, shop-fabricated, galvanized	Length	m	80	0.55	0.00	0.84
Duct work, shop-fabricated, stainless steel	Length	m	150	0.55	0.00	0.44
Evaporator (forced circulation)	Area	m^2	160,000	0.7	0.41	0.35
Filter (plates & press)	Area	m^2	6,800	0.58	0.26	0.42
Filter (rotary drum)	Area	m^2	32,000	0.63	0.26	0.27
Hoppers (conical)	Volume	m^3	58	0.68	0.00	0.04
Mills (ball)	Mass flow rate	tph	3,000	0.65	0.27	0.34
Mills (hammer)	Mass flow rate	tph	2,800	0.85	0.27	0.34
Screens (vibrating)	Area	m^2	18,000	0.58	0.12	0.18
Storage tanks	Volume	liters	870	0.3	0.20	0.23

Source: From reference [1] except for unit cost and sizing exponent for dryers, which are adapted from reference [3]; all data adjusted to 2002 dollars.

*Special materials of construction, operation at elevated pressures, and other factors may increase these unit costs.

$$C_P(1 \times 10^6 \text{ kg/hr}) = C_P(1 \text{ kg/hr})\left(\frac{1 \times 10^6}{1}\right)^{0.5}$$

$$= \$4557 \times 1000$$

$$= \$4.6 \text{ million}$$

The replacement pump must be able to provide 1×10^6 kg/hr of liquid water flow pressurized to 100 kPa. Notice from Table 10.12 that the sizing parameter for pumps with drivers is the product of volumetric flow rate (in cubic meters per minute) and pressure head (in kilopascal). For the specified pump, the sizing parameter is:

$$\frac{\text{volumetric}}{\text{flow rate}} \times \frac{\text{pressure}}{\text{head}} = \frac{1 \times 10^6 \text{ kg/hr}}{1 \times 10^3 \text{ kg/m}^3}\left(\frac{1 \text{ hr}}{60 \text{ min}}\right) \times 100 \text{ kPa}$$

$$= 1667 \text{ m}^3 \text{ kPa/min}$$

For this pump and driver the unit cost is $2007 and the sizing exponent for this pump is 0.52; thus, equation. 10.14 yields the replacement price for the boiler pump to be:

$$C_P(1667 \text{ m}^3 \text{ kPa/min}) = C_P(1 \text{ m}^3 \text{ kPa/min})\left(\frac{1667}{1}\right)^{0.52}$$

$$= \$2007 \times 47.4$$

$$= \$95,100$$

Scaling relationships can also be applied to overall systems, but with diminishing accuracy as the system becomes more complex. For many energy and chemical process plants, a reasonable estimate for n is 0.6, which yields the so-called *sixth-tenth rule* [3].

The extensive cost charts found in the literature were developed from prices effective in a particular year. Inflation can greatly increase the cost of equipment over the course of a few years and must be accounted for in estimating current equipment costs. This can be done with the relationship:

$$C_{p,c} = C_{p,p}(I_c/I_p) \tag{10.15}$$

where $C_{p,c}$ = inflation-adjusted cost of equipment in current year
            $C_{p,p}$ = known cost of equipment in a previous year
            I_c = inflation index factor for current year
            I_p = inflation index factor for the previous year in which equipment cost is known

Inflation index factors are available for different categories of equipment vs. year from a variety of sources. The Marshall and Swift (M&S) equipment index can be obtained from recent issues of *Chemical Engineering* magazine or *The Oil and Gas Journal*. Alternatively, the Consumer Price Index prepared by the U.S. Bureau of

Table 10.13. Inflation index factor based on the Consumer Price Index as published by the U.S. Bureau of Labor Statistics*

Year	Inflation Index Factor	Year	Inflation Index Factor
1955	0.151	1984	0.587
1956	0.154	1985	0.608
1957	0.159	1986	0.619
1958	0.163	1987	0.641
1959	0.164	1988	0.668
1960	0.167	1989	0.700
1961	0.169	1990	0.738
1962	0.171	1991	0.769
1963	0.173	1992	0.792
1964	0.175	1993	0.816
1965	0.178	1994	0.837
1966	0.183	1995	0.861
1967	0.189	1996	0.886
1968	0.197	1997	0.906
1969	0.207	1998	0.920
1970	0.219	1999	0.941
1971	0.229	2000	0.972
1972	0.236	2001	1.000
1973	0.251	2002	1.018
1974	0.278	2003	1.042
1975	0.304	2004	1.067
1976	0.321	2005	1.093
1977	0.342	2006	1.119
1978	0.368	2007	1.147
1979	0.410	2008	1.175
1980	0.465	2009	1.204
1981	0.513	2010	1.233
1982	0.545	2011	1.262
1983	0.562	2012	1.293

Source: U.S. Bureau of Labor Statistics (http://www.bls.gov/cpi/home.htm).

* Inflation factors are normalized to the year 2001. Inflation factors beyond the year 2001 are projections.

Labor Statistics can be used as an estimate of inflation index factors for all classes of equipment, as given in Table 10.13.

Materials of Installation and Labor of Installation

Installation of equipment can require considerable materials and labor: concrete and steel for an installation pad and support structure; electric wiring and control panels for motors; and piping and valves for water, gas, and steam utilities are among some of the more common items. Counting up all these costs represents considerable effort. Fortunately, *module factors* have been determined for both materials and labor associated with installation of various kinds of industrial equipment [1,2].

The *materials module factor (MMF)* is defined as the ratio of the cost of materials, C_M, to install a particular piece of equipment to the cost of the equipment (f.o.b.):

$$MMF = C_M/C_P \tag{10.16}$$

The *labor module factor (LMF)* is defined as the ratio of the cost of labor, C_L, to install a particular piece of equipment to the combined cost of the installed equipment and materials used to perform the installation:

$$LMF = C_L/(C_P + C_M) \tag{10.17}$$

Table 10.12 tabulates materials module factors and labor module factors for several important classes of equipment.

Direct costs, C_D, of purchasing and installing equipment is the sum of the equipment cost, materials of installation costs, and labor costs:

$$C_D = C_P + C_M + C_L \tag{10.18}$$

> **Example:** Calculate the total cost of installing the low-pressure boiler of the previous example.
>
> The f.o.b. cost C_P of the boiler was estimated to be \$4.56 million in the previous example. From Table 10.12, the MMF for installation of this boiler is 0.19. Thus, the cost of materials C_M is:
>
> $$C_M = MMF \times C_p = 0.19 \times \$4,560,000 = \$866,000$$
>
> For the same boiler the LMF in Table 10.12 is 0.26. Thus, labor cost C_L is:
>
> $$C_L = LMF \times (C_p + C_M) = 0.26 \times (\$4,560,000 + \$866,000)$$
> $$= \$1,411,000$$
>
> Total direct cost to purchase and install the boiler is:
>
> $$C_D = C_P + C_M + C_L = \$4,560,000 + \$866,000 + \$1,411,000$$
> $$= \$6,837,000$$

Thus, the direct cost to acquire and install the boiler is 50% higher than the f.o.b. price of the boiler. Notice that other kinds of equipment listed in Table 10.12 can have significantly higher ratios of direct cost to f.o.b. price than for this example due to material module factors that can exceed unity and labor module factors that approach 0.5.

Some references do not break down the costs of materials and labor into separate module factors. Instead, a *direct cost factor (M&L)* and a *direct labor-to-direct materials ratio (L/M)* are provided. These are defined by the equations:

$$\text{M\&L} = (C_P + C_M + C_L)/C_P \tag{10.19}$$

$$\text{L/M} = C_L/(C_P + C_M) \tag{10.20}$$

Most of the entries in Table 10.12 were developed from direct cost factors and labor/materials ratios given in Reference 1.

10.3.1.2 Indirect Costs

Indirect costs are expenses associated with the construction of a plant or facility that cannot be characterized as equipment, materials, or labor. The three most common indirect costs are categorized as freight, insurance, and taxes; construction overhead; and engineering expenses. The combined cost of *freight, insurance, and taxes* C_{FIT} can be estimated to be 8% of total equipment costs ($0.08C_P$) for projects in the United States. *Construction overhead* cost, C_O, includes fringe benefits on labor (heath insurance, sick leave, vacation and holiday pay, retirement benefits); so-called *burden on payroll* (Social Security taxes, unemployment insurance, workmen's compensation); salary and benefits for supervisory personnel; rental of construction machinery and purchase of small tools; and site clean-up upon completion of a project. This cost can be estimated to be 70% of labor costs ($0.7C_L$). *Engineering expenses,* C_E, include salaries and benefits for design and project engineers, office expenses, and associated overhead. It can be estimated to be 15% of equipment and installation materials cost ($0.15[C_P + C_M]$).

Installed bare module costs, C_{BM}, represents all direct and indirect expenses associated with purchase and installation of a piece of equipment (that is, the bare module). Installed bare module costs for auxiliary facilities, like complete power plants or wastewater treatment plants, are sometimes tabulated. More frequently C_{BM} are not directly tabulated for auxilary facilities and these must be determined by estimating the expenses associated with equipment installation.

10.3.1.3 Other Costs

Two other costs associated with an engineering project are contingency and fee and auxiliary facilities. Contingency refers to unexpected expenses on a project (weather-related delays, construction errors, poor estimation of costs, etc.) and is estimated to be 15% of bare module costs ($0.15C_{BM}$). *Fee* is essentially profit for the company performing the project and is estimated to be 3% of bare module costs ($0.03C_{BM}$). Notice that a well-managed project would consume little of the contingency, which instead becomes profit for the construction company. Thus, contingency and fee are a combined category equal to $0.18C_{BM}$. Adding contingency and fee to bare module cost gives the *total module cost*, C_{TM}.

The *cost of auxiliary facilities*, C_{AF}, for administrative offices, laboratory, maintenance shop, and warehouses are appropriate for some kinds of projects. In these cases, total cost of auxiliary facilities will range between 23 and 37% of total module cost. As an average, one can assume C_{AF} is equal to $0.3C_{TM}$. The sum of total module cost and auxiliary facilities cost gives the *grass-roots capital* of the project, C_{GR}. Notice that the grass-roots capital can be 3–5 times the cost of equipment installed in a facility or plan.

10.3.2 Operating Costs

Once a plant is constructed, funds are required to operate the plant. Like capital costs, *operating costs* include direct costs and indirect costs. In addition, there are *capital charges*, which represent payments on loans secured to construct the plant. Operating costs are typically calculated on an annual basis. Once the total operating cost is determined ($/yr), it is divided by the *annual production output* (units/yr) to determine the *annual production* cost ($/unit).

These various costs are conveniently tabulated in a *Summary of Operating Costs*, as illustrated in Table 10.14. At the top of the table is listed the *fixed-capital cost* for the project, which is equal to the grass-roots capital determined in Table 10.11. Fixed capital represents money that could not be easily recovered once spent. In addition to fixed capital, a sum of money called the *working capital* must be invested to get the plant into operation. It represents inventory of raw material and finished product. For inexpensive inventory, or an expensive plant, working capital is approximately 10% of fixed capital. For expensive inventory, or an inexpensive plant, working capital is closer to 20% of fixed capital. *Total capital* is the sum of fixed and working capital.

Another important quantity is the *capacity factor f_0* of a plant. Most plants or installed equipment do not operate 24 hours per day or 365 days per year: output may not be required continuously, the plant may close down at night, or equipment may require frequent maintenance and repair. The *capacity factor* is simply the fraction of time a plant operates on an annual basis. It is important in calculating the annual production output of a plant and the total amount of raw materials and utilities consumed. It does not generally affect other production costs such as labor and capital charges since these must be paid regardless of whether the plant is being operated or not.

10.3.2.1 Direct Costs

Direct costs include raw materials, by-product credits, operating labor, utilities, maintenance and repairs, and operating supplies. *Raw materials* are the inputs to the process, such as coal for a power plant or raw ore to a steel mill. Once the cost

Table 10.14. Summary of operating costs

Fixed capital		Equal to grass-roots capital
Working capital		10–15% of fixed capital
Total capital		Sum of fixed capital and working capital
Plant capacity factor (f_0)		Fraction of year that plant or facility operates
Production output (units/yr)		Annual production in kilowatts, gallons, etc. (adjusted to account for capacity factor)
	Cost ($/yr)	
Direct		
Raw materials		Calculated as: C_R($/kg) $\times \dot{m}$ (kg/s) $\times 31.5 \times 10^6$ s/yr $\times f_0$
By-product credits		Value enclosed in parentheses and subtracted from other costs.
Operating labor		See Table 10.15
Supervisory labor		10–20% of operating labor.
Utilities		See Reference 7.
Maintenance & repairs		2–10% of fixed capital
Operating supplies		10–20% of maintenance & repairs
Laboratory charges		15% of operating labor
Patents and royalties		3% of the sum of other direct expenses
Direct subtotal		Sum of all direct operating expenses
Indirect & General Expenses		
Overhead		50–70% of the sum of operating labor, supervision, and maintenance & repair
Local taxes		1–2% of fixed capital
Insurance		0.4–1.0% of fixed capital
General expenses		15% of operating labor + 5% of direct expenses
Indirect subtotal		Sum of all indirect operating expenses
Annual capital charges		Annual payment of interest and principal on loan for total capital $C_{TC} i(1 + i)^n/[(1 + i)^n - 1]$
Annual operating cost		Sum of direct costs, indirect costs, and annual capital charge
Product cost ($/unit production)		Annual operating cost divided by annual production output

per unit mass is determined from suppliers, data from the process flow chart can be used to calculate the cost of each raw material:

$$\text{Raw material cost} = C_R \times \dot{m} \times 31.5 \times 10^6 \times f_0 \qquad (10.21)$$

where C_R is the unit cost of raw material ($/kg) and $\dot{m}$ is the feed rate (kg/s) of raw material into the plant. Notice that the raw material cost is proportional to the capacity factor for the plant. Many plants yield by-products during the production of a desired product; for example, fly ash from a biomass-fired power plant or distillers' dried grains and solubles (DDGS) from a corn dry-milling plant. These represent credits if they can be sold; *by-product cost* is usually enclosed in parentheses and should be subtracted from other costs when annual production cost is summed. Calculation is similar to that employed for raw material cost.

Operating labor represents wages for people who operate equipment in the plant. This quantity can be estimated from tabulations of "Operator Requirements" such as those in Table 10.15. Obviously, simple equipment or fully automated plants may have little operating labor associated with them on an annual basis. *Supervisory labor* includes managers and clerical staff at the facility; this cost can be estimated as 10–20% of operating labor.

Utilities represent such process inputs as electricity, natural gas, potable water, and steam. The cost per unit can be obtained from such references as the *Statistical Abstracts of the United States* [7]. Annual cost can be calculated in a manner similar to that employed for raw materials and should include capacity factor.

Maintenance and repairs can be expected even for highly reliable equipment and should be included as part of the operating costs. Typically, these will be 2–10% of

Table 10.15. Operator requirements for various types of process equipment

Generic Equipment Type	Operators per unit per shift
Air plants	1
Boilers	1
Cooling towers	1
Water demineralizers	0.5
Electric generating plants	3
Portable electric generating plants	0.5
Mechanical refrigeration units	0.5
Wastewater treatment plants	2
Water treatment plants	2
Conveyors	0.2
Crushers, mills, grinders	0.5–1
Evaporators	0.3
Vaporizers	0.05
Furnaces	0.5
Fans	0.05
Blowers and compressors	0.1–0.2
Gas-solids contacting equipment	0.1–0.3
Heat exchangers	0.1
Mixers	0.3
Reactors	0.5
Clarifiers and thickeners	0.2
Centrifugal separators and filters	0.05–0.2
Bag filters	0.2
Electrostatic precipitators	0.2
Rotary and belt filters	0.1
Plate and frame, shell and leaf filters	1
Expression equipment	0.2
Screens	0.05
Size-enlargement equipment	0.1–0.3

Source: Adapted from Reference [3].

fixed capital, the low end representing well-established, relatively simple equipment and processes, and the high end for unconventional or speculative processes.

Operating supplies include replaceable materials in a plant not accounted for as part of regular maintenance. They can be estimated as 10–20% of maintenance and repair costs.

Laboratory expenses represent quality control testing or other chemical and physical analyses to support the manufacturing process. These are estimated as 10–20% of operating labor.

Expenses for *patents and royalties* occur when a process is licensed from another organization. This fee is usually fixed in advance, but for estimating purposes can be taken as 3% of the sum of all other direct expenses. In some manufacturing processes, of course, this fee may not apply.

10.3.2.2 Indirect and General Expense Costs

Indirect costs include overhead, local taxes, and insurance. *Overhead* includes fringe benefits, Social Security taxes, unemployment insurance, and retirement funds for workers at a facility. It can be a significant fraction of total labor costs and is usually estimated as 50–70% of costs for labor, supervision, and maintenance and repair (which is mostly labor). *Local taxes* are property taxes and can be estimated to be 1–2% of fixed capital. *Insurance* is 0.4–1% of fixed capital. *General expenses* arise from corporate activities associated with the operation of a plant. These include *administrative expenses*, taken as 15% of operating labor, and *distribution and marketing*, taken as 5–10% of total production costs.

10.3.2.3 Capital Costs

Typically, a loan of capital (both fixed and working) must be secured to build and start up a plant. We shall assume that a loan is secured for the total capital, C_{TC}, required to build and start up a plant. The annual interest rate of the loan is i (expressed as a decimal fraction) and the payment period of the loan is n years. To pay off the interest and principal of this loan will require the inclusion of *annual capital charges*, C_{CC}, in the operating costs equal to:

$$C_{CC} = C_{TC}\, i(1 + i)^n/[(1 + i)^n - 1] \tag{10.22}$$

The cost of capital can be a significant fraction of the cost of operating a facility. Depending on interest rate and the number of years of the loan, annual capital charges can run to 10–20% of the total capital and may dominate operating costs.

As an alternative to calculating capital charges, the plant can be considered as an investment by the providers of capital (stockholders), and must then generate a reasonable rate of return on the invested capital. In this case, a *selling price* is established to achieve an expected *return on investment (ROI)*, and a decision is made whether this price will be competitive in the open market. However, this analysis requires a sophisticated analysis of profitability, which is beyond the scope of this presentation. Instead, we calculate a simple production cost.

10.3.2.4 Annual Operating Cost and Product Cost

Once direct, indirect, and capital charges are calculated, they can be summed to give the *annual operating cost* of a plant. The *product cost per unit of production* is simply the annual operating cost divided by the annual production output of the plant. This number can be compared to product cost for competing processes to get an idea whether the plant or facility is worth pursuing. However, the question of whether a company can make money on the investment requires a more sophisticated analysis by qualified economists and marketing experts.

10.3.3 Detailed Example of Estimating Costs for Manufacturing a Biobased Chemical

As a detailed example of cost estimating, the production of ethanol in continuously stirred tank reactors is presented based on the analysis of Maiorella et al. [8]. A flow diagram for this ethanol production system is illustrated in Fig. 10.3. Molasses (50 wt-% glucose) and nutrients are mixed with water to the desired feed concentration and sterilized by direct steam injection in a continuous sterilizer. The sterilized solution is continuously added to five stirred fermentors, which are sparged with filtered air to maintain the optimum oxygen concentration. An absorber recovers vaporized ethanol from the fermentor vent gas.

The whole beer product is centrifuged to remove cell mass (yeast cream), which is fed to a rotary steam tube dryer operated with 600 psig steam to produce a dried yeast cattle feed supplement. The ethanol-water vapors released from the yeast are condensed in a stillage evaporator to provide a portion of the evaporation heat. The condensate is mixed with the supernatant from the centrifuge. The stillage evaporator generates some atmospheric pressure steam that represents an energy credit in the cost analysis.

The mixture of supernatant from the centrifuge and condensate from the stillage evaporator are fed to a distillation system consisting of a stripper and a vacuum column for ethanol recovery. The bottom liquid from the stripper contains residual cells, non-metabolized feed components, and fermentation by-product, including proteins. This bottom product is concentrated to 30 wt-% solids in a multi-effect evaporator and mixed back with the yeast cream feed to the rotary dryer as a further nutrient supplement. Thus, all waste streams are eliminated and no special waste disposal is required.

Plant capacity is taken to be 100 million liters per year of 95 wt-% azeotropic ethanol. The mass flows (kg/hr) of water, sugar, yeast, and ethanol in the plant corresponding to this capacity are illustrated in Fig. 10.3. The first step in estimating cost of production is to determine the cost of equipment to be installed in

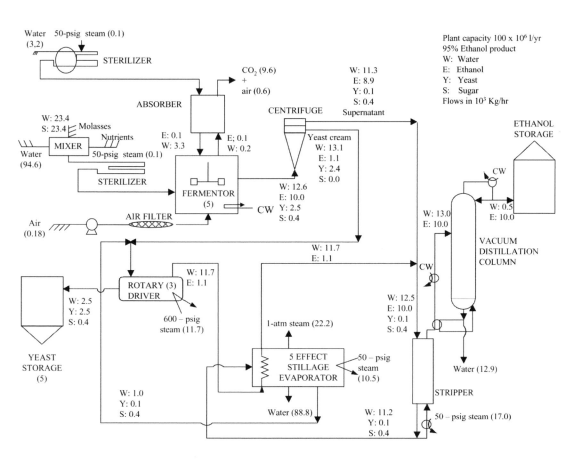

FIG. 10.3. Flow diagram for CSTR fermentor plant (adapted from B. L. Maiorell, H. W. Blanch and C. R. Wilke, 1984. *Biotech. & Bioeng.* 26:1003–1025).

the plant. Maiorella et al. [8] have determined equipment costs, C_p, for the proposed plant as summarized in Table 10.16 (adjusted to 2002 dollars). Total equipment costs are $10,800,000, which is only a fraction of the cost of installing the equipment in the plant. The costs for materials of installation and labor of installation are calculated from materials module factor MMF and labor module factor LMF, respectively, which were estimated from Table 10.12, as previously described. These module factors and the resulting materials and labor costs for installation are listed in Table 10.16 for each piece of equipment. Total cost of materials of installation is $4,840,000 while total cost of labor for installation is $5,740,000.

These numbers are the basis for preparing a summary of capital costs for the continuous fermentation ethanol plant as given in Table 10.17. Total direct expenses (the sum of costs for equipment, materials of installation, and labor of

Table 10.16. Equipment costs, materials of installation, and direct labor for installation of equipment for a continuous fermentation ethanol production plant

Item	Description	Equipment Cost C_P	MMF	Materials for Installation C_M	LMF	Direct Labor C_L
Storage (2 weeks)						
Molasses	7.86×10^6 L, carbon steel	$532,000	0.20	106,000	0.23	171,000
Ethanol	4.24×10^6 L, carbon steel	331,000	0.20	66,200	0.23	91,400
Yeast	3.23×10^5 L, carbon steel, 5 units	490,000	0.20	98,000	0.23	135,000
Screw conveyor for yeast delivery	0.40 hp, 100 ft long	16,000	0.27	4,320	0.25	4,000
Fermentation						
Fermentor	1.77×10^5 L, stainless steel, 5 units	880,000	1.04	915,000	0.49	880,000
Fermentor agitator	117 hp, stainless steel, 5 units	667,000	0.28	187,000	0.27	231,000
Fermentor cooler	41 m^2, stainless steel, 5 units	42,000	0.72	30,200	0.37	26,700
Air compressor	27.6 hp, 30 psig, 792 kg/h air	82,000	0.60	49,200	0.39	51,200
Feed-water sterilizer	Insulated stainless steel pipes plus heat exchangers, 1.23×10^5 kg/h water	179,000	0.72	129,000	0.37	114,000
Feed mixing tank	1.03×10^4 L, stainless steel	33,000	1.04	34,300	0.49	33,000
Centrifuge	237 hp, 1.26×10^5 L/h feed	679,000	0.28	190,000	0.23	200,000
Ethanol recovery						
Absorber	1 in. Raschig rings, gas rate = 1.02×10^4 kg/h	130,000	1.04	135,000	0.49	130,000
Absorber-water sterilizer	3340 kg/h water	25,000	0.72	18,000	0.37	15,900
Stillage evaporator	3790 m^2, 5 effects	3,314,000	0.41	1,359,000	0.35	1,635,000
Rotary dryer	667 m^2, 27.6 hp, 1.17×10^4 kg/h water removed, 3 units	2,893,000	0.28	810,000	0.36	1,333,000
Distillation	1.25×10^5 kg/h feed, feed EtOH wt % = 7.4, stripper followed by distillation column	685,000	1.04	712,000	0.49	685,000
Total		10,798,000		4,843,000		5,736,000

Sources: Equipment specifications and capital costs from Reference [8], adjusted to 2002 dollars. MMF and LMF values from Table 10.12.

Table 10.17. Summary of capital costs for continuous fermentation ethanol production plant

	Cost ($ millions)	Calculation
Direct project expenses		
Equipment (f.o.b.)	10.8*	C_P
Materials for installation	4.84*	$C_M = MMF \times C_P$
Direct labor	5.74*	$C_L = LMF \times (C_P + C_M)$
Total direct	21.4	$C_D = C_P + C_M + C_L$
Indirect project expenses		
Freight, insurance, taxes	0.864	$C_{FIT} = 0.08 C_P$
Construction overhead	3.99	$C_O = 0.7 C_L$
Engineering expenses	2.34	$C_E = 0.15(C_P + C_M)$
Total indirect	7.19	$C_{ID} = C_{FIT} + C_O + C_E$
Bare module cost	28.6	$C_{BM} = C_D + C_{ID}$
Contingency & fee	5.15	$C_{CF} = 0.18 C_{BM}$
Total module cost	33.8	$C_{TM} = C_{BM} + C_{CF}$
Auxiliary facilities	10.1	$C_{AF} = 0.3 C_{TM}$
Grassroots capital	43.9	$C_{GR} = C_{TM} + C_{AF}$

* From last row of Table 10.16.

installation) are $21,400,000. Indirect expenses (freight, insurance, taxes; construction overhead; and engineering expenses) are readily estimated as fractions of C_P, C_M, and C_L, respectively, as described in the notes accompanying Table 10.17. Bare module cost, the sum of direct and indirect expenses, comes to $28,600,000. Contingency and fee is assumed to add 18% to plant costs. Auxiliary facilities will likely be required for a plant of this size, which adds an additional 30% to plant costs, resulting in a grassroots-capital requirement of $43,900,000, which is a factor of 4.1 greater than the equipment cost alone.

Sometimes the grassroots capital is divided by the annual capacity of the plant to yield a capital cost per unit output, which in this case is $0.44 per liter of *annual ethanol capacity*. However, this number can be misleading as it suggests that capital costs are directly proportional to plant capacity, whereas the concept of economies of scale indicates that this capital cost per unit output gets smaller as the plant gets larger. Neither should this number be confused with the production costs per unit of output, which requires a calculation of annual operating costs.

A summary of operating costs for this continuous fermentation plant is given in Table 10.18. Raw materials include nutrient solution, water, and molasses. Utilities include electric power, cooling water, and both low-pressure and high-pressure steam. Credits are obtained on the yeast recovered and sold as cattle feed supplement and on steam from the stillage evaporator. Operating labor is based on 24 hour per day operations (3 shifts) working 330 days per year. Annual capital charges are based on a 20-year loan at 10% annual interest rate on total capital. Other expenses were based on percentages of fixed capital and operating labor as detailed in the notes included in Table 10.18. Total production costs are estimated at $50,450,000 per year, which represents a unit production cost of $0.505 per

Table 10.18. Summary of operating costs for a continuous fermentation ethanol plant

Fixed Capital	43.9×10^6	Working Capital	4.4×10^6	Total Capital	48.3×10^6
Plant capacity factor	0.9	Plant capacity	100×10^6 liters/yr		
	Cost (10^6/yr)	Description			

	Cost (10^6/yr)	Description
Direct		
Raw materials		
Nutrient solution	1.11	$0.11/1000 liter ethanol product
Water	0.109	$0.13/1000 liter
Molasses	34.70	$85/ton (50 wt% sugar, 95% utilizable)
Raw materials subtotal	35.9	
By-product credits		
Yeast credit	(4.81)	$0.144/kg of 50 wt% yeast
Steam credit (1 atm pressure)	(0.293)	$1.67/1000 kg
By-product credits subtotal	(5.10)	
Operating labor	1.27	$15/man-hr, 10 workers per shift, 3 shifts per day
Supervisory labor	0.190	15% of operating labor
Labor subtotal	1.46	
Utilities		
Power	0.616	6.98×10^6 kW-h/yr @ $0.088/kW-h
Cooling water	0.264	$0.06/1000 liter, 20°C
Low pressure steam (340 kPa)	1.79	$6.93/1000 kg
High pressure steam (4080 kPa)	0.901	$9.71/1000 kg
Utilities subtotal	3.57	
Maintenance & repairs	2.63	6% of fixed capital
Operating supplies	0.395	15% of maintenance & repairs
Laboratory charges	0.191	15% of operating labor
Direct subtotal	39.0	Sum of all direct operating expenses
Indirect and general expenses		
Overhead	2.45	60% of total labor & maintenance & repair
Local taxes	0.878	2% of fixed capital
Insurance	0.307	0.7% of fixed capital
General expenses	2.14	15% of operating labor + 5% of direct expenses
Indirect subtotal	5.78	Sum of all indirect operating expenses
Annual capital charges	5.67	10% interest for 20 years on total capital
Annual operating cost	50.45	Direct costs + indirect costs + capital charge
Product cost ($/liter)	0.505	Operating cost divided by production output

Source: Reference [8].

liter. This number is about twice the current production cost of ethanol from well-run starch-to-ethanol plants in the United States. However, it is in line with ethanol costs in 1984 when Maiorella et al. performed the analysis [8].

10.4 Costs of Manufacturing Various Biobased Products

Electricity from Combustion of Biomass [9]

The capital and operating costs for steam power plants fired with biomass are relatively well known because of significant operating experience with these systems.

The capital cost for a new plant ranges between $1,400 and $1,800 per kilowatt capacity. Accordingly, a 50-MW biomass power plant based on direct-combustion would cost approximately $80 million. On the basis of a target price of $1.90/GJ for biomass, the cost of production for direct-fired biomass power is about $0.06/kWh in 1990 dollars.

Electricity from Gasification of Biomass [10–12]

The capital cost for a gasification plant, including fuel feeding and gas clean up, is dependent on both the size and the operating pressure of the system. An atmospheric-pressure gasifier producing 50 MW of thermal energy would cost about $15 million. A comparably sized pressurized gasifier would cost $57 million because of the added complexity of lock hoppers and pressure vessel design. Pressurized systems improve gasification rates for coal. However, biomass is so reactive that there is little chemical kinetic or thermodynamic advantage in pressurized gasification of biomass. The producer gas from a utility-scale biomass gasifier can be used in either a gas turbine or fuel cell to produce electricity.

A gasification/gas-turbine power plant producing 50 MW of electricity would have total capital cost of between $75 million and $138 million (between $1500 and $2750/kW), the smaller number reflecting improved technical know-how after building at least ten plants. Electricity production costs would range from $0.05 to $0.09/kWh if fuel is available at an optimistic price range of $1.00–$1.50/GJ.

Capital costs for high temperature fuel cells suitable for integrated gasification/fuel cell power plants currently cost $3000/kW. Molten carbonate fuel cells are expected to be $1500/kW at the time of market entry but decrease to about $1000/kW for a commercially mature unit. The cost of electricity from a mature unit operating on natural gas is projected to be between $0.049 and $0.085 per kWh. More attractive economics result if less expensive fuel is available. The cost of electricity generated from landfill gas using mature fuel-cell technology is expected to be comparable to that for an internal combustion engine/electric generator set, i.e., about $0.05/kWh.

Biogas from Anaerobic Digestion [13]

Anaerobic digestion is commercially developed, but for the purpose of treating wastewater rather than for power production. The methane generated is often fired in internal combustion engines to produce electricity in an effort to help offset costs of waste treatment, but does not have immediate prospects for replacing natural gas. Capital costs for anaerobic-digestion facilities processing more than 200 tpd of volatile solids are estimated to be between $44,000 and $132,000 for each ton per day of volatile solids processed. Methane yields will be approximately 0.38

m³/kg of volatile solids. Thus, a 200 tpd anaerobic-digestion plant could produce 28 million cubic meters of methane per year, representing almost 2900 GJ/day of chemical energy. Projected operating costs for producing methane from dedicated energy crops are in the range of $5–$6/GJ in 1986 dollars. In comparison, the cost of natural gas in the United States, which shows large seasonal and geographical variations, ranges between $1.90 and $4/GJ. In niche markets, where the feedstock is inexpensive and natural gas is not available, biogas can be a viable alternative energy resource.

Ethanol from Biomass [14,15]

The cost of producing ethanol from biomass varies tremendously depending on the feedstock employed, the size and management of the facility, and the market value of co-products generated as part of some conversion processes. Industrial-scale ethanol plants resemble petrochemical plants in that they are capital intensive and benefit from economies of scale.

Cost information for ethanol plants to be built in the United States is most reliable for those using cornstarch, the basis of the U.S. ethanol industry. A 5,000 barrel per day plant (about 265 million liters per year) built from the ground up will have a capital cost of about $140 million in 1987 dollars, or $0.53/liter of annual capacity. Smaller facilities can have capital costs as high as $0.79/liter of annual capacity, and poorly designed facilities of any size may cost $1.06/liter of annual capacity. On the other hand, ethanol plants that are built from existing facilities such as refineries or chemical plants, or ethanol plants integrated into a larger industrial facility, can have substantially lower capital costs, often in the range of $0.26–$0.40/liter of annual capacity.

Low-end production costs are about $0.26/liter. However, the volumetric heating value of ethanol is only 66% that of gasoline. This production cost, therefore, is equivalent to gasoline selling for $0.39/liter before tax, transportation, or profit. In contrast, refinery price for gasoline in 1990 dollars was about $0.20/liter. Currently, the economics of fermentation are such that the commercial viability of ethanol is entirely dependent on government incentives in the form of a tax credit, currently $0.16 for each liter of ethanol used for fuel blending. Also, a strong market for fermentation by-products is a key factor in the economic viability of ethanol-from-corn.

Technology to convert lignocellulose to sugar is expected to reduce the cost of fuel ethanol, although detailed economic information is not currently available. Capital cost for a 5,000 barrel per day plant to produce ethanol from lignocellulose using simultaneous saccharification and fermentation (SSF) is estimated to be $175 million (1994 dollars). Assuming wood costs $42/dry ton, ethanol can be produced for about $0.31/liter. Combining economies of scale with advances in processing technology are projected to decrease production costs to $0.13/liter.

However, some reports suggest that ethanol from cellulose will have to cost as little as $0.08–$0.11/liter to be competitive with the gasoline prices anticipated early in the twenty-first century.

Methanol from Biomass-derived Syngas [16]

Capital investment for a 7,500 barrel per day plant to produce methanol from biomass would be about $280 million in 1991 dollars. The cost of methanol from $40/dry ton of wood is projected to be about $0.27/liter. Since the volumetric heating value of methanol is only 49% that of gasoline, the production cost from this plant is equivalent to gasoline selling for $0.55/liter. Methanol from natural gas can be produced at significantly lower cost, but this assumes much larger plant capacities to capture economies of scale. Such large plants are not feasible for widely dispersed biomass feedstocks. New methanol-synthesis technologies may be able to significantly reduce this price. The U.S. Department of Energy's methanol from biomass program has a goal of $0.15/liter ($7.90/GJ) based on feedstock cost of $1.90/GJ.

Bio-oil from Fast Pyrolysis [17]

Capital investment for a 5000 barrel per day plant to produce pyrolysis liquids would be $63 million in 1987 dollars. Assuming biomass feedstock was available at $1.70/GJ, this size plant could produce pyrolysis liquids for $0.18/liter, which has an energy value of $6.70/GJ.

Biodiesel from Vegetable Oil [18]

Capital costs for a biodiesel facility are relatively modest, costing about $250,000 for a 50 barrel per day plant (3.2 million liters per year). However, feedstock costs for production of biodiesel are relatively higher than feedstocks for production of other kinds of fuel, ranging from $0.16 to $0.26/liter for waste fats to $0.53–$0.79/liter for vegetable oils. Under the best scenario, a biodiesel plant might produce fuel for $0.44/liter. Diesel fuel produced from petroleum typically sells for less than $0.25/liter.

Further Reading

1. Guthrie, K. M. 1969. "Data and Techniques for Preliminary Capital Cost Estimating." *Chem. Eng* (March 24): 114–142.
2. Guthrie, K. M. 1974. *Process Plant Estimating, Evaluation, and Control.* Solano Beach, Calif.: Craftsman.

3. Ulrich, G.D. 1984. *A Guide to Chemical Engineering Process Design and Economics*. New York: Wiley.
4. Perry, J. H. and C. H. Chilton. 1973. *Chemical Engineers' Handbook*, 5th Edition. New York: McGraw-Hill.
5. Peters, M. S. and K. D. Timmerhaus. 1980. *Plant Design and Economics for Chemical Engineers*, 3rd ed. New York: McGraw-Hill.
6. Boehm, R. F. 1987. *Design Analysis of Thermal Systems*. New York: Wiley.
7. U.S. Department of Energy. 2000. *Statistical Abstracts of the United States for 2000*.
8. Maiorella, B. L., H. W. Blanch and C. R. Wilke. 1984. "Economic evaluation of alternative ethanol fermentation processes." *Biotechnology and Bioengineering* 26:1003–1025.
9. Environmental Law & Policy Center. 2001. *Repowering the Midwest: The Clean Energy Development Plan for the Heartland* (available on the World Wide Web: www.repowermidwest.org/ repoweringthemidwest.pdfx)
10. Bridgwater, A. V. 1995. "The Technical and Economic Feasibility of Biomass Gasification for Power Generation." *Fuel* 74:631–653.
11. Williams, R.H. and E. D. Larson. 1993. "Advances in Gasification-based Biomass Power Generation." *Renewable Energy: Sources for Fuels and Electricity.* T. B. Johnson, H. Kelly, A.K.N. Reddy and R. H. Williams, eds. Washington, D.C.:Island Press.
12. Hirschenhoofer, J. H., D. B. Stauffer and R. R. Engleman. 1994. *Fuel Cells: A Handbook*. Department of Energy Technical Report DOE/METC-94/1006 (DE94004072).
13. Benson, P. H., T. D. Hayes and R. Isaacson. 1986. "Regional and Community Approaches to Methane from Biomass and Waste: an Industry Perspective." In proceedings *Energy from Biomass and Wastes X*. Orlando, FL.
14. National Advisory Panel on Cost Effectiveness of Fuel Ethanol Production. 1987. *Fuel Ethanol Cost-Effectiveness Study Final Report*.
15. Lynd, L. R., R. T. Elander and C. E. Wyman. 1996. "Likely Features and Costs of Mature Biomass Ethanol Technology." *Applied Biochemistry and Biotechnology* 57/58:741–761.
16. Klass, D. L. 1998. "Thermal Conversion." Chap. 10 in *Biomass for Renewable Energy, Fuels, and Chemicals*. New York:Academic Press.
17. Elliott, D. C., A. Ostman, S. Borje Gevert, D. Beckman, Y. Solanantausta, C. Hornell and B. Kjellstrom. 1990. In "Energy from Biomass and Wastes XIII." D. L. Klass, ed. Chicago IL: Institute of Gas Technology.
18. Gavett, E. E., D. Van Dyne and M. Blasé. 1993. In "Energy from Biomass and Wastes XIII." D. L. Klass, ed. Chicago: Institute of Gas Technology.

Problems

10.1 A production system for soybeans consists of plowing with a chisel plow, followed by a field cultivator, and finally a planter. Harvest consists of combining (with a head for beans) and hauling the beans on the farm. If the yield is 40 bushels per acre, what are the direct and indirect costs associated with the use of machinery in this operation?

10.2 A hypothetical genetically modified corn is developed with a yield of 5 Mg/ha/yr that fixes nitrogen in the soil. Production costs are similar to those for conventional corn (see Table 10.7), except that seed is 25% more expensive and no nitrogen fertilizer is required. What is the unit production cost ($/Mg)? Could this corn be grown profitably at today's market price for corn?

10.3 A short-rotation woody crop has a market price of $35/ton. In the first year, capital costs are $600/ha to purchase land, and expenses are $400/ha to plant and fertilize the crop. Annual expenses in years 2 through 6 are only $50/ha. In the seventh year, when the woody crop is harvested, expenses increase to $550/ha. The harvest yields 75 tons/ha. Use a simplified cash-flow analysis to determine whether this enterprise is profitable. Assume a 10% discount rate.

10.4 A chemical plant for the production of 5000 barrels per day of pyrolysis oils was estimated to cost $63 million in 1987.

(a) Estimate the cost of a plant with twice this capacity in 1987.
(b) Estimate the cost of a plant of 5000 barrel per day capacity in 2003.

10.5 A plant requires a new feedstock preparation system to chop lignocellulosic material into fibers at a rate of 50 tons per hour. The system consists of hammermill and a 0.6 m-wide conveyer belt that moves the chopped material a distance of 10 m. Estimate the grassroots-capital requirement for this installation.

10.6 A plant designed to produce 50 barrels per day of biodiesel has a grassroots-capital requirement of $250,000. Direct operating costs (materials, labor, utilities, and maintenance) are $3.5 million/yr and indirect operating costs (overhead, taxes, insurance, and general expenses) are $1.5 million/yr. Capital charges are based on 10% interest over 10 years. What is the production cost of biodiesel ($/liter)?

Appendix

Conversion Factors

Area

1 hectare (ha) = 10,000 m^2
1 hectare (ha) = 2.471 acres
1 acre = 43,560 ft^2

Volume

1 liter (L) = 10^{-3} m^3
1 liter (L) = 0.2642 gallons
1 cubic meter (m^3) = 35.31 cubic feet
1 cubic meter = 264.2 gallons
1 barrel (bbl) = 42 gallons
1 bushel (bu) = 1.244 ft^3
1 bushel corn = 25.4 kg
1 bushel wheat = 27.2 kg
1 bushel soybean = 27.2 kg

Weight

1 kg = 2.205 lb
1 ton (metric) = 1000 kg
1 ton (English) = 2000 lb
1 ton (metric) = 1.1 ton (English)
1 ton per day (tpd) = 0.015 kg per sec (kg/s)

Energy

1 joule (J) = 1 kg m^2/s^2
1 kilojoule (kJ) = 10^3 J
1 megajoule (MJ) = 10^6 J
1 gigajoule (GJ) = 10^9 J

1 terajoule (TJ) = 10^{12} J
1 petajoule (PJ) = 10^{15} J
1 exajoule (EJ) = 10^{18} J

1 British thermal unit (Btu) = 1054 J
1 thousand Btu (MBtu) = 10^3 Btu
1 million Btu (MMBtu) = 10^6 Btu
1 quadrillion Btu (quad) = 10^{15} Btu
1 quad = 1054×10^{15} J (approximately 1 EJ)

1 toe (metric ton oil equivalent)
   = 7.4 barrels of crude oil in primary energy
   = 7.8 barrels in total final consumption
   = 1270 m^3 of natural gas
   = 2.3 metric ton of coal

Power

1 kilowatt (kW) = 1 kJ/s
1 megawatt (MW) = 1000 kW
1 MW = 3.415 MMBtu/hr
1 horsepower (hp) = 2546 Btu/hr
1 horsepower (hp) = 550 ft-lb/s

Index

Note: Boldface numbers indicate illustrations and tables.